*How We are Lied to, Cheated and Manipulated
by Statistics...and Why You Should Care*

SCAMMED

BY STATISTICS

Edward Zaccaro
Daniel Zaccaro

Ed lives outside of Dubuque, Iowa, with his wife Sara. He has been involved in various areas of education since graduating from Oberlin College in 1974. Ed holds a Masters degree in gifted education from the University of Northern Iowa and has presented at state and national conferences in the areas of mathematics and gifted education.

Daniel Zaccaro is currently a graduate student at the University of Northern Iowa studying psychology.

Art Work: Jack Berg, Galena, Illinois

Layout and design by Blue Room Productions (memories2movies@aol.com)

Hickory Grove Press, 3151 Treeco Lane, Bellevue, Iowa 52031.

Phone: 563-583-4767
E-mail: challengemath@aol.com
www.challengemath.com
Library of Congress Control Number: 2010912026
ISBN 10: 0-9679915-7-9
ISBN 13: 978-0-9679915-7-3

Books by Edward Zaccaro:

Primary Grade Challenge Math

*Challenge Math for the Elementary and
Middle School Student*

Real World Algebra

*The Ten Things All Future Mathematicians and
Scientists Must Know (But are Rarely Taught)*

Becoming a Problem Solving Genius

*25 Real Life Math Investigations
That Will Astound Teachers and Students*

*Scammed by Statistics
How We are Lied to, Cheated and Manipulated by Statistics*

This book is dedicated to my father,
Luke N. Zaccaro (1924 - 1977)
a mathematician, critical thinker and philosopher who
instilled in his children a love of life,
reason and responsibility.

"Statistical thinking will one day be as necessary for efficient citizenship as the ability to read and write."

--- H. G. Wells

Table of Contents

INTRODUCTION

"Mathematics is how the universe talks
to us and reveals its truths."
--- Galileo

"Mathematics is a more powerful instrument of
knowledge than any other that has been
bequeathed to us by human agency."
--- Descartes

"Numbers are the highest degree of knowledge.
It is knowledge itself."
--- Plato

The beauty of statistics, and what makes them so powerful, is that they are cruelly indifferent to our hopes, dreams and beliefs --- they give us an objective look at a situation. Unfortunately, statistics are often treated like referees and umpires that can be argued with and manipulated when we don't like what they tell us. As the following example highlights, it can be very dangerous to ignore their message.

In 1999, a large pharmaceutical company was undergoing the final stages of testing for a blockbuster painkilling drug called Vioxx. Because Vioxx provided pain relief without the gastrointestinal complications associated with aspirin, it had the potential of not only helping hundreds of thousands of people, but also of making billions of dollars for its manufacturer.

The pharmaceutical company knew that the final testing for Vioxx needed to be done carefully --- especially the selection of the competition for the clinical trial. After extensive research and deliberation, it was decided that the "competition" for the trial would be Aleve (naproxen). (Aleve was picked because it was not known to have a protective effect against heart attacks.)

After nine months, the data from the clinical trial was analyzed and the results were shocking! The group that took Vioxx had four times the number of heart attacks as the Aleve group. The message in the statistics was very clear --- **There was a very strong possibility that Vioxx was a powerful cause of heart attacks.**

Unfortunately, the people who interpret statistics often are unable or unwilling to look at them objectively. They are susceptible to being in-fluenced by incompetence, wishful thinking and greed. So instead of acknowledging that Vioxx increased the risk of cardiac events by 400%, the claim was made that Aleve reduced the risk of cardiac events by 80%. This was an extraordinarily implausible interpretation because Aleve, as mentioned earlier, was not known to have a heart protective benefit like aspirin. In fact, if Aleve did reduce the risk of heart attacks by 80%, it would be two or three times more effective than aspirin!

Even though the statistics from the clinical trial made it clear that Vioxx was dangerous, the FDA gave its approval and Vioxx was subsequently used by millions of people. Vioxx was eventually pulled from the mar-ket four years later, but not before it caused a staggering number of heart attacks and deaths. The FDA estimated that Vioxx caused be-tween 88,000 and 139,000 heart attacks --- 30 to 40 percent of which were fatal.[1]

There was a clear message in the statistics from the Vioxx/ Aleve study. The message was ignored and tens of thousands of people died.

"Mathematics is how the universe talks to us and reveals its truths."

"Mathematics is a more powerful instrument of knowledge than any other that has been bequeathed to us by human agency."

"Numbers are the highest degree of knowledge. It is knowledge itself."

When Galileo, Descartes and Plato spoke those words, they clearly understood the extraordinary power of mathematics. Our society has had this power available to us for hundreds of years --- a power that when used properly, has the potential to save millions of lives. Unfortunately, there have been countless times when "statistical warnings" have been distorted, manipulated, and minimized. The result of this intellectual and moral failure has been the needless loss of millions of lives.

We took far too long to heed the warnings statistics gave us about the following:

- Tobacco

- Asbestos

- Benzene

- Vioxx

- Propulsid

- Lead

- Reyes syndrome / aspirin connection

- Alcohol

The numerous tragedies caused by the misuse and manipulation of statistics should not lead us to believe that statistics are always manipulated, useless, or never to be trusted. For every incident where statistics are used improperly, there are a hundred cases where they are used in a fair and reasonable way --- with great benefit to society. The five examples below illustrate the beneficial uses of statistics:

- A statistical model helped prevent over 100,000 deaths due to hospital error over an 18 month period.

- The Oakland A's used statistics to become one of the best teams in baseball, while paying players some of the lowest salaries in baseball.

- A mathematical formula proved to be more accurate at predicting the quality of wine than expert wine tasters.

- A statistical model was able to predict Supreme Court votes more accurately than a group of nationally known law experts.

- Statistics are used to help emergency room doctors make better decisions.

The ability of statistics to improve our lives makes it essential that we know how to use them. In addition, because manipulation, deception and outright lying often accompany the use of statistics, it is imperative that we know how to interpret statistics and are aware of the numerous techniques people use to distort and misuse numbers.

As you make your way through this book, you will be amused by some of the examples and outraged by others. My hope is that after you finish, you will not only know how to question the statistics you see, but also to see that the study of statistics is not the dry and boring endeavor it is often portrayed to be.

CHAPTER ONE
Barely Believable Graphs

"Lottery: A tax on people who are bad at math."

--- Author Unknown

The Cholesterol Lowering Power of Oats

Cholesterol is an essential fatty substance that is found in human cells and blood. Although the body cannot function without cholesterol, people with high cholesterol levels may develop fatty deposits in their blood vessels, which can lead to heart attacks and strokes.

Cholesterol in the blood comes from two sources: diet and the liver. Because of this, high cholesterol is usually treated with a combination of diet, exercise and medication. The current thinking in the medical community (2010) is that total cholesterol should be under 200 mg/dl.

One of the foods currently marketed as a fairly painless and healthy way to lower cholesterol levels is oats. One large food company used a bar graph in an advertising campaign to help people understand the cholesterol lowering power of oats. The y-axis has been covered in the bar graph below so you can guess how much oats lowered cholesterol levels in the four-week study.

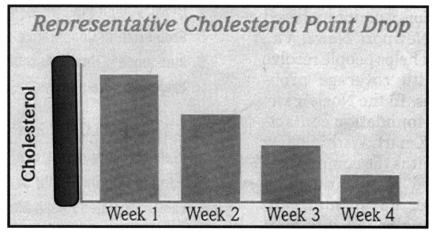

The visual appearance of the bar graph gives the impression that after four weeks of eating oats, a typical cholesterol level will drop by approximately 75%, which is good news if 1) you have an extremely high cholesterol level and 2) the bar graph really is a fair portrayal of the real drop in cholesterol. If your cholesterol level was at 400, you might think that it will drop to 100 in four weeks --- just by adding oats to your diet.

Let's uncover the y-axis and see if the bar graph is a fair representation of the effect of oats on cholesterol.

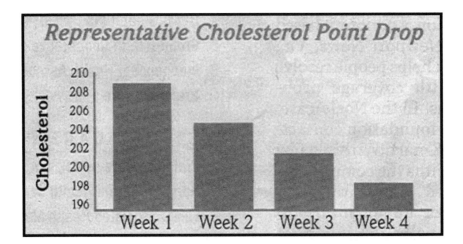

Instead of a 75% drop in cholesterol level as the visual implies, the scale tells the true story --- a measly 4% drop! Of course a 4% drop is better than nothing, but we can put this 4% drop in perspective by comparing it to the effect that a placebo had on cholesterol in a study that tested the cholesterol lowering power of Crestor (a statin).

During the study, the control group took a placebo while the test group took Crestor. The results were very interesting --- Crestor lowered cholesterol by 35 to 60 percent and the placebo lowered cholesterol by 7%!

I guess the message is that eating oats is helpful in lowering cholesterol, but if you want to lower it even more, use a placebo.

The statistical deception of the cholesterol lowering power of oats was not limited to print advertising. A radio ad by the same company, touting the cholesterol lowering power of oats, obviously could not show a bar graph.

"In the study, oats lowered cholesterol by 4%. If you don't think that 4% is a lot, try paying 4% more of your income in taxes!"

Noooo!!!!!!!

This comparison is not only meaningless, but it is clearly meant to deceive. You could just have easily have said: If you don't think that 4% is a lot:

- "Try fitting 4% of the population of New York City in your home!"

- "Try paying off 4% of the national debt!"

- "Try lifting 4% of the earth's weight!"

If you want to know why the company used this type of deception in advertising the cholesterol lowering power of oats, look at the fair graph below.

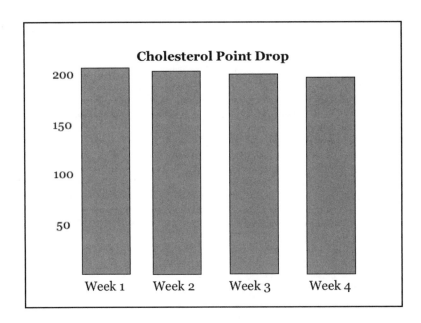

(A graph with a scale starting at 100 would also be fair because cholesterol usually does not go below 100.)

Making the Crime Rate Look Better While it Gets Worse
(Fictional Story)

A city was frustrated by its high murder rate, so the city council hired a new police chief to see if better leadership would reduce the murder rate. The number of murders that had occurred in the previous four years are shown below:

1990....................200 murders
1991....................205 murders
1992....................195 murders
1993....................200 murders

We need changes and we need them now!
The murder rate is awful!
This calls for a new chief!

The new police chief was hired at the end of 1993 and he agreed to return four years later to show the city council that he was having a positive impact on the crime rate.

At the end of 1997, the police chief was gathering statistics so he could report his progress to the city council. The council was grumbling because the local newspaper reported that there appeared to be more criminal activity in the city. Unfortunately for the police chief, the number of murders had significantly increased. The year after he arrived, murders went from 200 to 300. The next year, they jumped to 375 and the third year they increased to 425. In the fourth year they reached the 450 mark.

There must be a way to present these numbers to make me look good.

1994...................300 murders	
1995...................375 murders	
1996...................425 murders	
1997...................450 murders	

The chief decided to present the statistics in graphic form to the city council. Look carefully at the title of the bar graph.

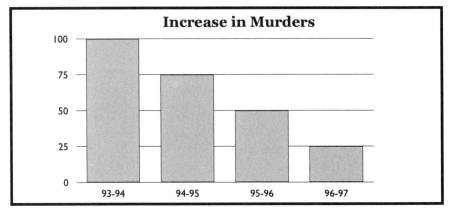

Increase in Murders

	93-94	94-95	95-96	96-97

Keep up the good work!
Impressive graph!

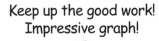

I am proud to report a dramatic decline in the increase in murders each year.

I don't know what it is, but something fishy is going on.

The city council was impressed with the chart and congratulated the chief on the impressive decline in the yearly **increase of murders**. They asked to be updated four years later. At the end of 2002, the police chief again had a problem. The number of murders continued to go up, and this time the amount of increase was even climbing.

1998.....................550 murders
1999.....................655 murders
2000.....................765 murders
2001.....................880 murders

This time even the amount of murders increased each year! There must be some way of looking at this in a positive way.

I present to you today some strong evidence that the percent of increase in murders is steadily declining.

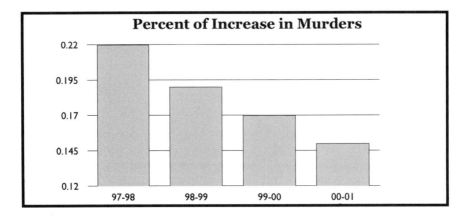

Percent of Increase in Murders

Looks like a good steady drop!
As long as the graph is going down.
I feel good about the crime rate.

We are going to hit
bottom soon and
I think that's good!

These guys are
starting to scare me.

The police chief didn't lie to the city council, but he gave such a distorted picture of the truth that it wasn't much different from lying.
This story is a dramatic example of how statistics can be manipulated to show almost anything you want. The message of this story
isn't that you shouldn't pay attention to statistics. The lesson is that
you must analyze statistics very carefully.

Which Car Company is Better?

One way to measure how well a company makes cars is to see what percent of the cars are still being driven after 10 years. (Story is based on actual advertising in the early 1980's)

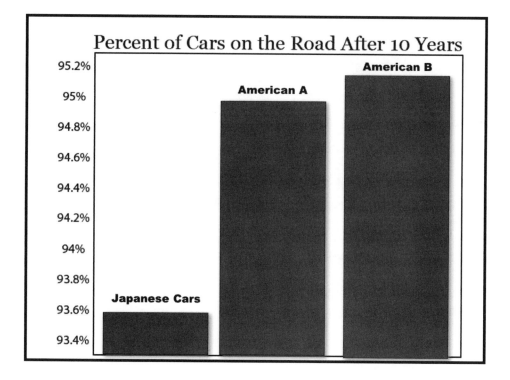

When you look at this graph, it is easy to think that the quality of American cars far exceeds the quality of Japanese cars. Make sure you look at the scale when you look at bar graphs. In this graph, all three car manufacturers have nearly the same percentage of cars on the road after ten years, but the graph makes it appear that the quality of the American cars far exceeds the quality of the Japanese cars. If I were a Japanese car manufacturer, I would show you a graph like the one on the next page.

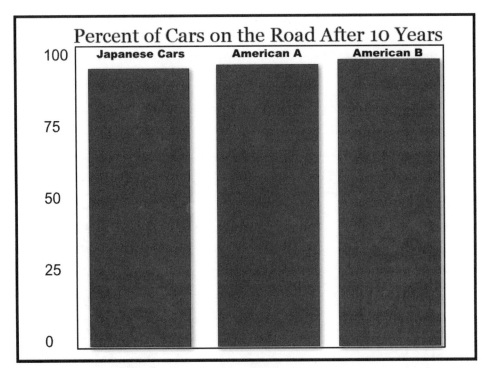

Percent of Cars on the Road After 10 Years

Now it is clear that there is very little difference between the car companies. Remember that bar graphs are visuals and they can be easily misinterpreted if one does not look at the scale and intervals. Both graphs use the same information, but present it quite differently.

Creating Partisan Divide With a Bar Graph

On February 25, 1990, Michael Schiavo found his wife, Terri, suffering from cardiac arrest. Shortly afterwards, a lack of oxygen to her brain caused her to lapse into a vegetative state, leaving her with reflexes such as eye blinking but depriving her of any cognitive ability. This set the stage for the seven year legal battle between Terri Schiavo's parents, who insisted that her feeding tube remain operating, and her husband, who after years of care had given up hope of recovery. (The majority of medical experts believe that the odds of recovering from a persistent vegetative state are essentially zero.)

The legal battle soon gained publicity as several activist groups joined the side of Terri's parents and established funds to pay for their legal battle to keep Terri Schiavo alive. In 2005, after three court rulings and despite significant opposition from the Governor of Florida and other conservative lawmakers, Terri Schiavo's feeding tube was removed and she died several days later.

The media maintained close scrutiny throughout the legal struggle surrounding Terri Schiavo, and both political sides rallied support from their bases. As a result, CNN.com reported the results of a poll asking if Democrats, Republicans, and Independents "agreed with the court's decision to have the feeding tube removed." This is the graphic CNN.com used to display the results:

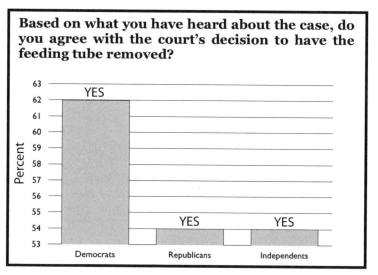

The impression the bar graph gives is that Republicans and Independents did not agree with the court's decision to have Terri Schiavo's feeding tube removed, while a large number of Democrats did agree with the decision. In fact, although the bar for the Democrats has 9 times the area of the other two bars, the disparity between the three parties is a mere 8 percentage points. Because the poll's margin of error is described as +/- 7 percentage points, the results indicate that the majority of Americans agree with the court's decision to remove Terri's feeding tube despite differences in political affiliation.

To their credit, CNN.com revised its visual after reports of the original graphic's deception surfaced. The updated graph correctly displays the scale of the results, but mistakes such as these cannot be reversed with updated visuals, as a large number of people had already viewed - and been influenced by - the deceptive original graph.

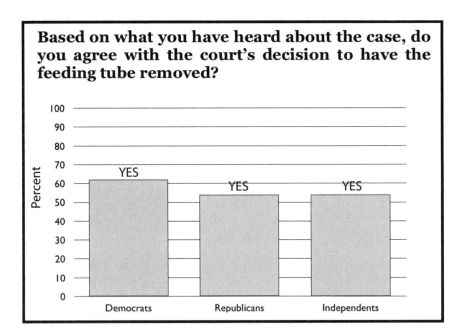

The Online Ad Recession is Officially Here

Similar bar graph scaling deception was used by Erick Schonfeld in a May 1st, 2009 article written for Techcrunch.com titled "The Online Ad Recession is Officially Here: First Quarterly Decline in Revenues."

In the article, Mr. Schonfeld uses a bar graph to show quarterly reports from the top four internet advertising companies (Google, Yahoo, Microsoft, and AOL).

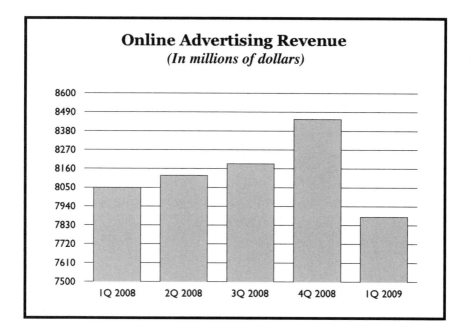

Just as in the CNN.com graph about Terri Schiavo, this graph does not show the true scale of the results because the y-axis begins at a number significantly higher than zero. Although the visual decrease from the 4th quarter of 2008 to the 1st quarter of 2009 appears to be over 50%, the numbers tell a different tale. The drop from the 4th quarter of 2008 to the first quarter of 2009 is only 589 million, a mere 7% drop from the previous quarter, not the staggering 50% decrease which the graph visually implies.

A revised graph shows the true scale of internet advertising revenue:

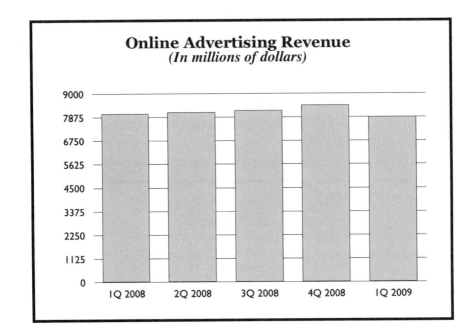

The new bar graph shows that although online advertising revenue did see a small decrease in early 2009, the online advertising business is far from bankrupt.

Another interesting aspect of the deceptive online revenue bar graph is that the graph did not show any data earlier than the year 2008. This clever deception hides the fact that advertising revenue for the last two quarters of 2007 was actually lower than the so-called recessive 1st quarter of 2009. Maintaining a scale similar to Schonfeld's original graph, but extending the time period by two quarters, we can see that his argument would not be supported had he revealed just two more data points. (Which is probably why he conveniently left them out of his visual.)

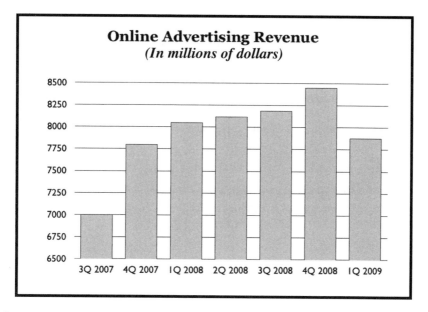

When people objected to the deceptive graph, the following "corrected" line graph replaced the bar graph:

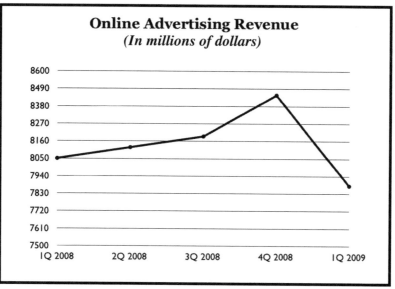

Schonfeld released the following statement:
"For everyone complaining in comments about my top chart because it doesn't start at zero, here is another one showing the exact same data."

Obviously using a poorly-scaled line chart is no different from using a poorly-scaled bar graph. For comparison, here's what the updated line chart should have looked like:

Online Advertising Revenue
(In millions of dollars)

The two bar graphs on the following page show how easy it is to present data that supports your point of view whether it is that online advertising is affected by the recession or it is not affected by the recession.

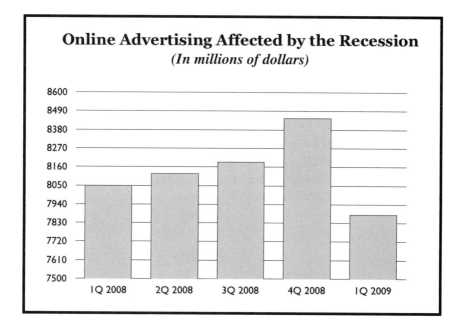

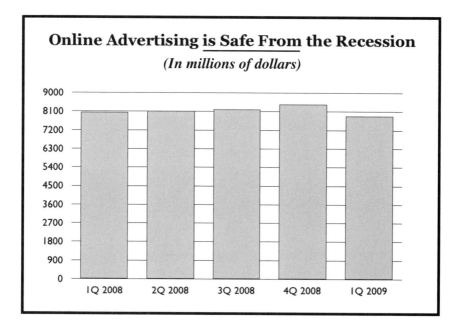

Making SAT Scores Look Better Than They Are

In addition to starting the y-axis at a number other than zero, another way to make a deceptive graph is to remove values for the y-axis completely, as the Oklahoma State Department of Education did on its website touting the state's high SAT scores:

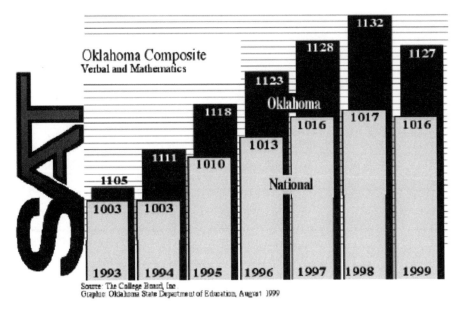

(These graphs were constructed by the Oklahoma State Board of Education using statistics provided by The College Board.)

Looking at the state's graphic, it is impossible to tell where the y-axis begins. This graph is also confusing for another reason: it is completely irrelevant to the actual numbers. In fact, the graph does not adhere to any scale at all, and the space between the horizontal lines does not represent a consistent number. Look at the difference between Oklahoma and the national average for the year 1993. We can clearly see that there is a difference of 102 points, which takes the space of two horizontal lines. From this year we can derive that each space stands for 51 points.

Moving to the year 1994, the difference between national and state scores is 108. For this year, Oklahoma scores rise 8 spaces over the national score. But wait...if each y-axis spacing stood for 51 points, as the previous year dictates, it would mean that Oklahoma's average composite scores for 1994 would equal 1411, much higher than the actual score of 1111. It seems that the creators of this graph used two separate scales for the national and Oklahoma average SAT scores, which is much more deceptive than simply beginning the y-axis at a number greater than zero (or in this case, 400, which was the lowest possible SAT score for these years). To better illustrate the point, here is a table constructed from the numbers from this graph:

National Score	Oklahoma Score	Difference Between Oklahoma and National Score	Number of Horizontal Spacings Between Scores	**Number of Points Represented per Spacing**
1003	1105	102	2	**51**
1003	1111	108	8	**13.5**
1010	1118	108	8	**13.5**
1013	1123	110	10	**11**
1016	1128	112	12	**9.3**
1017	1132	115	15	**7.7**
1016	1127	111	11	**10.1**

Oklahoma created their bar graph in a very interesting way. The difference between national and state scores is not represented consistently. If you look at ONLY the national scores or ONLY the state scores, each space **does** represent a consistent value of one SAT point. For example, national scores went up seven points from 1994-1995, which is accurately represented by seven y-axis spaces. The only way to make a graph such as the one Oklahoma created is to create two separate bar graphs, each with different y-axis scales, and superimpose them over each other. The problem with this approach is that IT'S WRONG!!! There is no way to visualize the difference between the national and state scores with the way Oklahoma presented its graph, which is exactly what visuals such as bar graphs should do. Here is what the graph should have looked like:

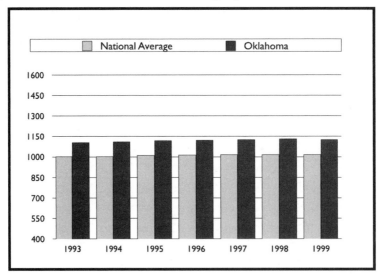

Using an honest scale, it is easy to see that Oklahoma's scores really are slightly higher than the national average. This in itself would be impressive enough to display on their website, but unfortunately they resorted to deceptive visuals.

Fortunately, the original graph can no longer be found on Oklahoma's State Department of Education website (the graph is 10 years old, after all), but misleading charts like this provide a testament to how careful we must be when interpreting graphs and scales.

Skyrocketing Consumer Prices for
Heating Fuels in the Midwest

A newspaper article showed four line graphs from The Energy Information Administration that highlighted the large increase in the cost of heating fuels over an eight year period starting with the winter of 2000-2001. While the graphs gave the appearance of equally dramatic price increases in each type of heating fuel, the numbers told a different story.

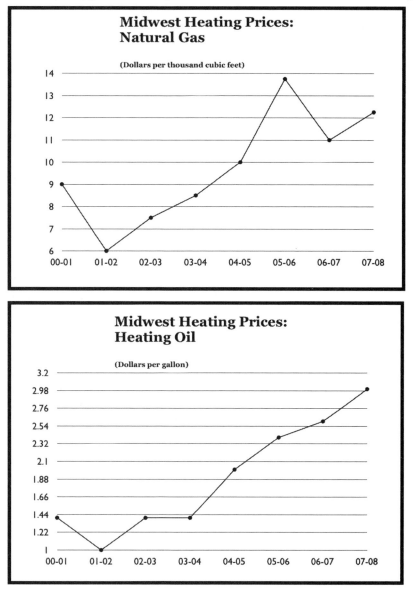

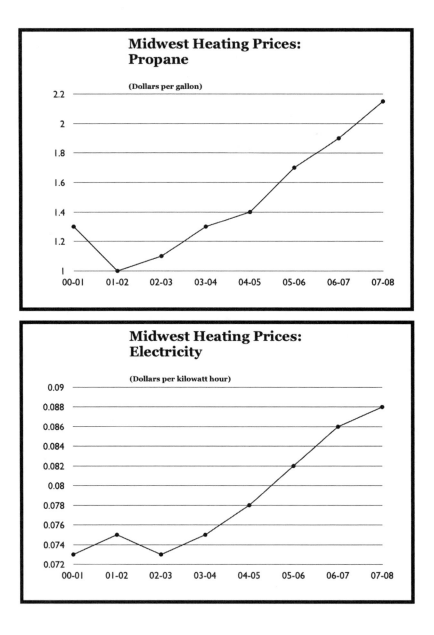

Let's look at the numbers carefully to determine the percent of increase for each fuel.

Natural Gas
Lowest Price: $6 per thousand cubic feet
Highest Price: $13.75 per thousand cubic feet

Percent increase: 130%

Heating Oil
Lowest Price: $1 per gallon
Highest Price: $3 per gallon

Percent increase: 200%

Propane
Lowest Price: $1 per gallon
Highest Price: $2.15 per gallon

Percent increase: 115%

Electricity
Lowest Price: $.74 per kilowatt hour
Highest Price: $.87 per kilowatt

Percent increase: 17.5%

The graphs should have allowed consumers to make quick comparisons among the four types of fuels. Unfortunately, because the price of all four heating fuels appeared to increase equally, many consumers undoubtedly missed the fact that the price of fuel oil went up 200%, while the price of electricity remained remarkably stable.

The graph below provides a more accurate representation of the actual increases in fuel costs.

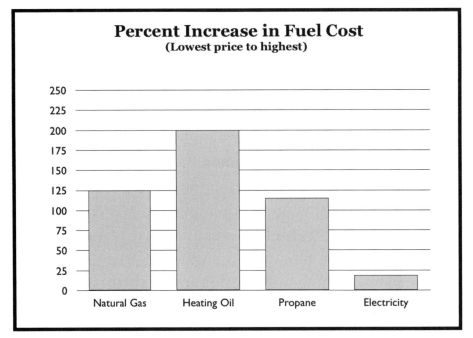

Are Traffic Fatalities Dropping?

A newspaper article (Dubuque Telegraph-Herald 5/21/05) that discussed traffic deaths in four area counties used a line graph that could easily have left many readers with the inaccurate perception that traffic fatalities in all four counties dropped dramatically in 2005.

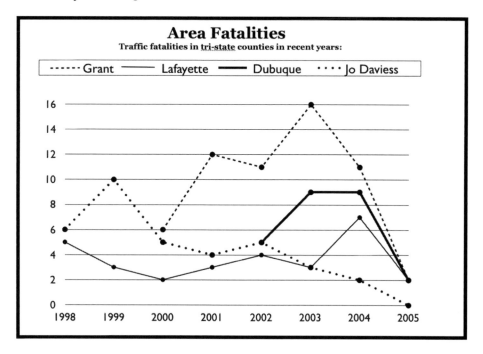

The reason the number of traffic fatalities appears to plummet in 2005 is because that year includes traffic deaths for only 4 months (January - April), while all other years include statistics for a full 12 months.

A fair graph would not have included statistics for 2005.

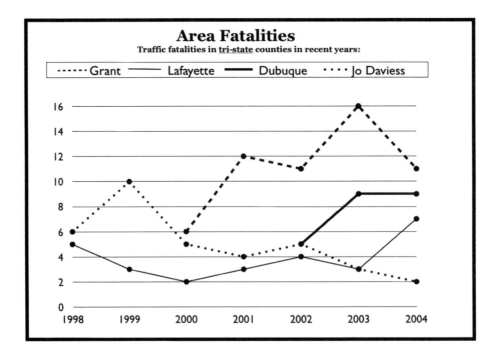

Making the Housing Slump Appear Worse Than It Is

Sales of existing homes started declining in early 2007 because of an economic slump and more restrictive loan standards. Even though the decline was significant, the National Association of Realtors made the situation appear much worse than it actually was by starting the y-axis in their graph at 5.5 million instead of zero.

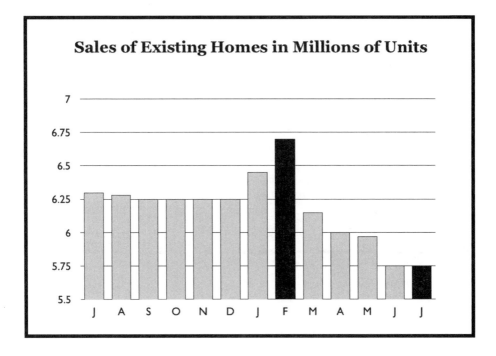

The seasonally adjusted annualized rate of sales for February 2007 was 6,700,000 homes while sales in July were 5,750,000 homes. This is a drop of 14%. The graph gives the appearance of an 80% drop in sales!

A more honest portrayal of the sales slump, one which uses zero as the beginning of the y-axis, would be a graph similar to the one shown below.

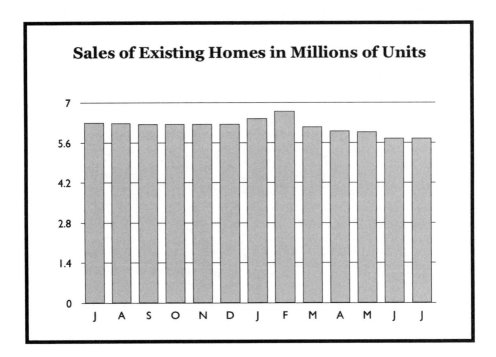

Giant Gold Panda Coins

The use of deceptive visual information is not limited to graphs. Advertisers often use pictures that lead many consumers to believe that an item being sold is much larger than it actually is. Look at the following picture of a gold coin that has been advertised in magazines for $129 + shipping and handling.

Prices and availability subject to change without notice. Actual coin is 13.92 mm.

Unless you read the fine print and understand how to convert from millimeters to inches, you will not be aware of the actual size of the coin you are ordering.

After seeing the large coin in the ad, it is very likely that those who order the gold panda coin would expect to receive a coin the size of a silver dollar, half dollar, or at least the size of a quarter. What they receive is a gold coin slightly larger than an aspirin! Look below to see the actual size of the gold panda coin that cost over $129.

The actual size of the gold coin is 13.92 mm in diameter, which is approximately 1/2 inch. This makes the area of the coin in the advertisement 65 times larger than the actual coin.

The Subtle Deception of Dropped Ceiling Bar Graphs

The following bar graph was presented by school administrators to the Dubuque Community School Board in 1998. It shows the academic ranking of Dubuque students compared to the United States as a whole. Notice that the scale ends at the 70th percentile instead of the 99th percentile.

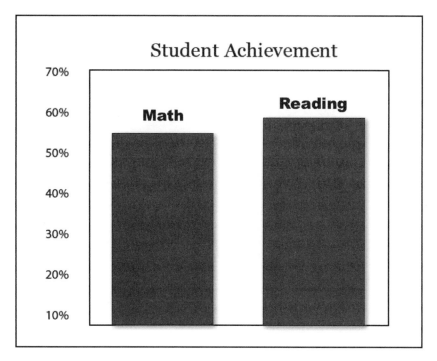

In this case, a dropped ceiling bar graph gives the visual impression of higher academic achievement because the bars are much closer to the "ceiling" than they would be if a fair scale was used.

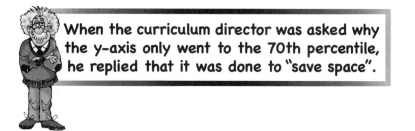

When the curriculum director was asked why the y-axis only went to the 70th percentile, he replied that it was done to "save space".

If a student received poor grades on his or her final exams, the scores could be presented deceptively by using a dropped ceiling bar graph.

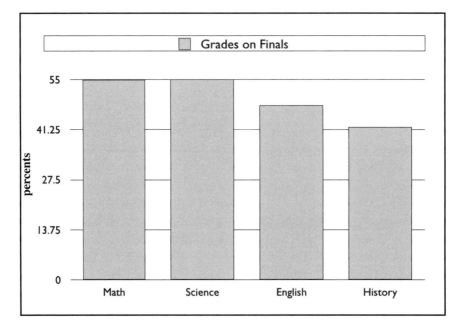

If the student decided to present his or her scores fairly :

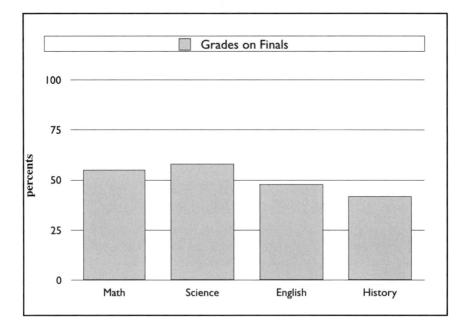

CHAPTER TWO

Are You Comparing Equivalent Groups?

Is California More Dangerous Than Iraq?

On August 26, 2003, television commentator Brit Hume attempted to put some perspective on the number of U.S. soldiers who were dying in Iraq:

"Two hundred seventy-seven U.S. soldiers have now died in Iraq, which means that statistically speaking, U.S. soldiers have less of a chance of dying from all causes in Iraq than citizens have of being murdered in California, which is roughly the same geographic size. The most recent statistics indicate California has more than 2300 homicides each year, which means about 6.6 murders each day. Meanwhile, U.S. troops have been in Iraq for 160 days, which means they're incurring about 1.7 deaths, including illness and accidents each day."

I was going to vacation in California. Now I think I will go to Iraq instead.

The fact that California and Iraq are "roughly the same geographic size" is meaningless in this situation because California has a population of 38,000,000 and the number of American soldiers in Iraq at the time was 170,000. When you look at the statistics fairly, it is very clear that Iraq is a much more dangerous place to be than California.

U.S. soldiers in Iraq: 170,000
Deaths in the 160 days from the beginning of the war: 277
Death rate per 100,000: 163 deaths

Population of California: 38,000,000
Murders per year: 2300
Murders in 160 days: 1008
Murder rate per 100,000: 2.7 murders

The probability of an American soldier dying in Iraq was approximately 60 times higher than the probability of being murdered in California. Brit Hume's attempt to equate death rates in Iraq and California because of their similar geographic size is as meaningless as saying:

U.S. soldiers have less of a chance of dying from all causes in Iraq than citizens have of being murdered in California which are similar because:

• **they both have the letters I, R, and A in their names** •

When another reporter pointed out Mr. Hume's misuse of statistics, Hume's response was: "Admittedly, it was a crude comparison, but it was illustrative of something."

 This guy is really starting to scare me.

The point that Brit Hume was trying to make when he gave the American public the false impression that soldiers were in no more danger in Iraq than they would be walking the streets of California was not that 277 American deaths was not a terrible tragedy. The point that he should have made was that American casualties in Iraq were negligible by historical standards. (American deaths in the Civil War, WWI, WW II, Korea, and Vietnam were substantially higher.)

Global Warming and the Frightening Increase
in Hurricane Damage

In Al Gore's thought provoking documentary "An Inconvenient Truth", he presented several statistics that highlighted current problems that he felt were the result of climate change. Many of Vice-President Gore's concerns are certainly legitimate and he has strong support from a large part of the scientific community, but his claim that the cost of hurricane damage had dramatically increased due to climate change is clearly a distortion.

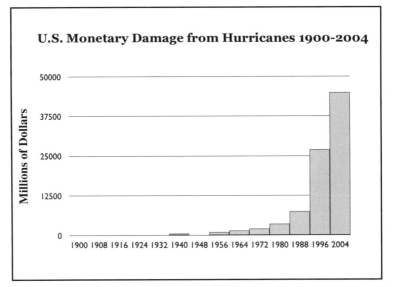

Graph based on data from the book *Physics for Future Presidents*

The data Gore uses in "An Inconvenient Truth" showing increased property damage caused by hurricanes is very misleading because it shows yearly storm damage without adjusting for inflation, population growth, and wealth. The dramatic inflation that has occurred over the past hundred years has given the appearance of very little damage in the earlier part of the century. (At the time, Sears was selling homes in its catalog for $2000 -$3000. A comparable home today would cost well over $100,000.)

In addition to inflation, the number of people who live on the coast has increased dramatically, along with the size and value of their homes. If we adjust for inflation, population growth and increased wealth, our graph of hurricane damage undergoes significant changes.

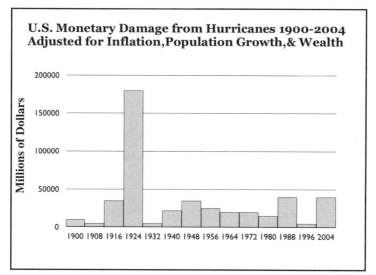

Graph based on data from the book *Physics for Future Presidents*

Because Al Gore distorted the impact of climate change on hurricane damage does not mean that those who doubt that climate change is a serious problem are right. Even though Vice-President Gore damaged his credibility by misusing statistics, there is still a significant amount of evidence that global warming is a serious problem. (A dramatic increase in property damage caused by hurricanes should not be considered part of that evidence.)

Disproving one position does not prove the opposing position or even necessarily give evidence for the opposing position.

If Al Gore wanted to present a fair graph that compared hurricane damage in the first half of the century to hurricane damage in the second half, he needed to keep his points of comparison as similar as possible for the 100 years his graph represented. When fair data is plotted, there is no apparent trend in hurricane damage --- up or down.

It is also very difficult to compare the number and strength of hurricanes from the year 1900 to the present. Our ability to detect and measure hurricanes has clearly changed over the last 100 years. Before detection buoys and satellites, a hurricane would have to hit shore or impact a shipping lane to be recorded.

If we try to make our yearly measurements consistent over a 100 year period, we should count only hurricanes that have reached the coast of the United States. When we do that, there is no trend up or down for the number or the strength of hurricanes.[2]

The Appearance of Dramatic Academic Achievement Gains in a Midwest City

Dubuque is a small picturesque city that sits on the banks of the Mississippi River. In the winter of 1994, Dubuque Community School administrators anxiously awaited the results of achievement tests that thousands of students had taken the previous fall. Many changes had occurred in the previous couple years and those who were involved were eager to see if there would be any improvement in student test scores. In addition to the hiring of a new superintendent, the district also made several other changes they hoped would have a positive impact on student test scores:

- **New school reforms had been implemented at several schools in the district.**

- **The test date was moved ahead several weeks.**

- **Teachers made a special effort to talk to students about test-taking strategies.**

When the test results came out, administrators could not help but smile --- the changes in student achievement were stunning! All eleven elementary schools showed dramatic gains. The graphs shown on the following page are for the 4th grade of each of Dubuque's elementary schools. (The 5th, 6th, 7th and 8th grades made similar achievement gains.)

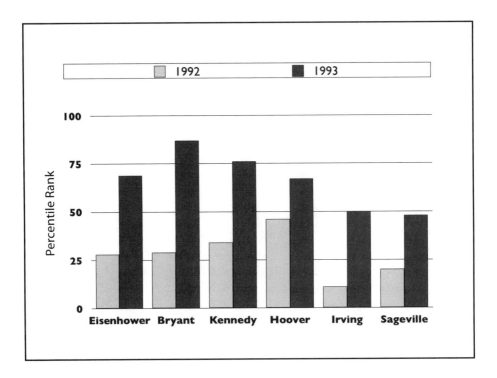

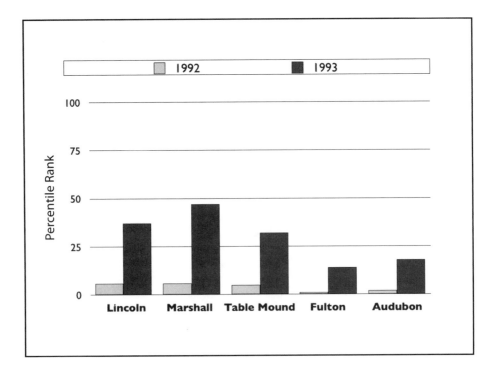

"Marshall school's 4th grade scores jumped from the 6th percentile in 1992 to the 47th percentile in 1993."

"What is especially exciting is that lower socioeconomic schools had dramatic achievement gains!"

"Audubon School is only at the 18th percentile, but compared to the previous year's 2nd percentile, this is very good news."

When there are achievement gains this dramatic, you should always be a little skeptical --- make that <u>very</u> skeptical!

The Dubuque Telegraph Herald (local newspaper) shared the good news with the community. In response to the very noticeable jump in scores, the new superintendent said: "They're very good. The principals and teachers and students and parents did a very good job as evidenced by the improvement of our rankings compared to our nation and to Iowa."

How could the Dubuque Community Schools have accomplished such unprecedented achievement gains when so many other cities struggled to squeeze out, at best, meager gains each year? The answer to that question was buried in the newspaper article.

TELEGRAPH HERALD
MARCH 6, 1994

"Some parents of special education students have complained that their students were tested separately from their classmates and that their scores were not included in the averages."

Nooooo!!!!!!!!

Approximately 10-15% of the lowest scoring students in the district were removed from the testing pool in 1993 and then test scores were compared to the previous year's test scores when the lower performing students were included in the test pool.

The dramatic gains in student achievement were not accomplished through a special curriculum or innovative teaching techniques or hard work. They were the result of statistical manipulation.

Let's look at the composite test scores of 4th graders in Dubuque's School District over the next few years. If Dubuque found the secret formula for academic success, the scores should continue to rise. If the perceived achievement gains were from test group manipulation, then we should see a jump from 1992 to 1993 when the manipulation occurred, and then see very little change in subsequent years.

The following graph shows that this is precisely what happened:

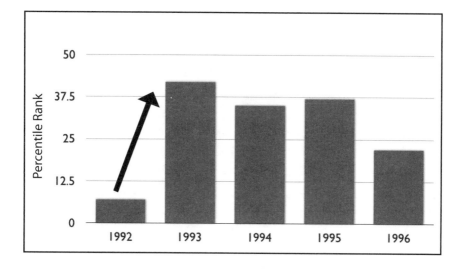

It is very important to keep test groups the same when undertaking a statistical analysis. In this case, the Dubuque School District did not. If the data schools compile on student achievement is not completely honest and if it is not an accurate assessment of the effectiveness of instructional practices at our schools, then we blind ourselves to the reality of the situation and can make unwarranted conclusions based on the misleading statistics.

In addition to manipulating test scores to make them look high, we also must not make the mistake of minimizing low scores by saying they are only a one day snapshot of a student's academic abilities.

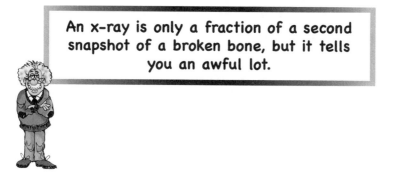

An x-ray is only a fraction of a second snapshot of a broken bone, but it tells you an awful lot.

Is a Rental Assistance Program Associated
With an Increasing Crime Rate?

For years, the city of Dubuque, Iowa has heard allegations that its participation in a Housing Choice Voucher Program, commonly known as Section 8 housing, has led to an increase in crime in the city.

> **Section 8 housing is a program run by the Department of Housing and Urban Development. Qualifying low-income participants can have up to 2/3 of their rent paid with a Section 8 housing voucher.**

The issue was even part of a local political campaign when one candidate called for a review of Section 8 housing as part of a crime prevention initiative. Many objected to the candidate's "insult" to Section 8 housing participants, including the local newspaper's editorial board, which called his comments "an allegation that plays on people's fears, reinforces biases and hasn't been proven."

Eventually the Dubuque Housing Commission, which administers the Section 8 housing program, hired a research firm to determine once and for all if Section 8 renters were involved in more criminal activity than other citizens of Dubuque.

The research firm decided to look at arrest reports for a designated two-month span of time when crime is typically at its highest level. They took all arrested individuals and put them into one of three categories:

(1) Renters from Section 8 housing
(2) Renters not involved with Section 8 housing
(3) Those who were not from Dubuque and those who lived in private homes

Representatives from the research firm presented their results to a relieved Housing Commission at their next meeting.

17% of those arrested lived in Section 8 housing

20% of those arrested lived in other rental units

63% of those arrested were non-Dubuque residents or lived in private homes

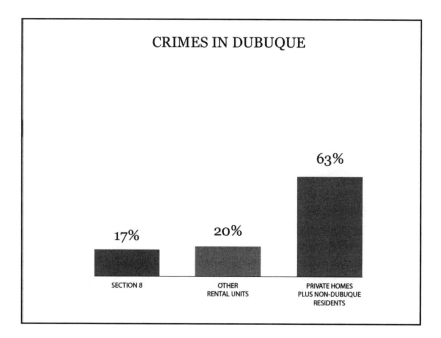

The research firm concluded that not only were Section 8 renters less likely to commit crimes, but in fact they had the lowest number of arrests of the three groups the researchers analyzed.

The Dubuque Telegraph Herald was about to go to print with the news that Section 8 renters had been unfairly tied to crime when an internal discussion at the paper raised questions about the study and the editor decided to pull the story at the last minute.

What the paper subsequently discovered was that the study looked at numbers of arrests and did not take into account that the three groups were not equivalent in size. In fact, the Section 8 participants were only 4% of the population of Dubuque and were responsible for approximately 20% of the criminal arrests during the two months the study took place.

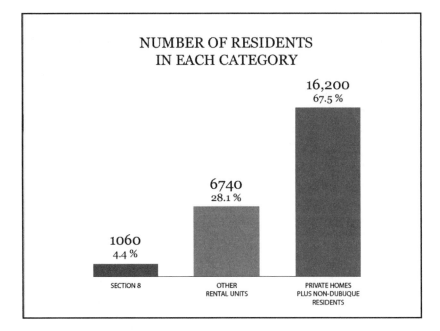

NUMBER OF RESIDENTS
IN EACH CATEGORY

16,200
67.5 %

6740
28.1 %

1060
4.4 %

SECTION 8

OTHER
RENTAL UNITS

PRIVATE HOMES
PLUS NON-DUBUQUE
RESIDENTS

When the size of each group is taken into account, it is clear that there is a strong link that connects Section 8 housing and crime. A look at arrests per 1000 residents, instead of total arrests in each group, makes the picture much clearer.

(1) Renters from Section 8 housing:
1060 with 121 arrests

(2) Renters not involved with Section 8 housing:
6740 with 136 arrests

(3) Residents of Dubuque who lived in private homes:
16,200 with 296 arrests

(142 nonresidents of Dubuque were arrested)

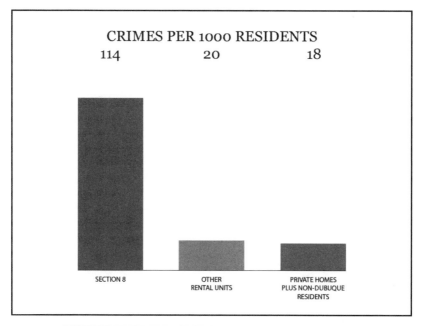

CRIMES PER 1000 RESIDENTS

| 114 | 20 | 18 |

| SECTION 8 | OTHER RENTAL UNITS | PRIVATE HOMES PLUS NON-DUBUQUE RESIDENTS |

The arrest rate for Section 8 housing was approximately six times the rate of other rental units and private homes.

The research firm clearly should have known that it was manipulating statistics when it presented raw numbers in the manner that it did. What if Dubuque and New York City each had 10 murders in a year's time? Could it be claimed that the chance of being murdered is equivalent in each city? Of course not. Dubuque has a population of 65,000 and New York City's population is over 8 million.

When the question of Section 8 housing's connection to crime arose, the city of Dubuque needed a reasoned discussion on the subject. An integral part of that discussion was a fair statistical analysis of the link between crime and Section 8 housing. If the misleading statistics that showed no connection went unchallenged by the local newspaper, a proper assessment and honest debate about the pros and cons of Section 8 housing would not have been possible.

CHAPTER THREE
Painting Bull's-eyes Around Arrows

This archer appears to be very skilled until you realize that he shot his arrows and then painted the targets. Some researchers use the same technique as the archer to make the results of their research fit their expectations. They cherry-pick data that reinforces their beliefs, instead of having their beliefs modified by the results of a study or experiment.

The Meditation Experiment

For two months in the summer of 1993, over 5000 practitioners of Transcendental Meditation came to Washington D.C. to take part in an experiment. Their objective was to prove that thousands of practicing meditators could dramatically lower the level of violence in a city by simply meditating in unison. (This would create a "coherent consciousness field.")

> This "coherent consciousness field" would not only lower the stress of the meditators, but would also spread out throughout the city and lower the level of violent crime in Washington D.C.

A young physicist named John Hagelin, who led the project, held a press conference to explain the meditation experiment and to outline its goals and objectives.

You will see a scientific demonstration that will provide proof of a unified superstring field.

When people use the phrase "unified superstring field", I tend to think that I am about to hear nonsense.

Mr. Hagelin's premise was that the large number of people meditating in unison would form a superstring field which would spread out throughout the city and promote peace and tranquility.

The scientific community was understandably skeptical of the violence-reduction meditation project, but they tried to keep an open mind as the two-month experiment began. Almost no one in the scientific community expected that any definitive results from the experiment would occur. They thought there would probably be just the normal fluctuations in the number of murders which take place in a large city over the summer months.

What did happen that summer was truly shocking! Contrary to the expectations of most outside observers, the results of the meditation project were definitive and unambiguous. The murder rate skyrocketed to an unprecedented level!

The project leader, while acknowledging that murders did increase, promised that he would soon release the results of the unusual experiment after all the data was scientifically analyzed.

A year later, a detailed report was released that strongly implied that the two-month meditation experiment had been an unqualified success. The report, which was reviewed by an "independent scientific review board", claimed success because their analysis showed the following:

• There was a significant reduction in psychiatric emergency calls.

• There were fewer complaints filed against the police in Washington D. C.

• President Clinton's approval rating went up during the two-month period.

The focus of the experiment was turned away from the murder rate, assaults and robberies. Instead, the experimenters picked anything positive from the data (which was really difficult) and ignored the rest. The data from the experiment was made to fit their beliefs, instead of having their beliefs modified by the new information.

An interesting postscript:

The "independent scientific review board" consisted entirely of practitioners of Transcendental Meditation.

Great News for Arthritis Patients --- or Maybe Not

In the fall of 2000, promising results from a major clinical trial of Celebrex showed that it reduced the incidence of ulcers over a six-month time period. This was exciting news for gastroenterologists, who were interested in new medications that did not cause stomach bleeding or ulcers.

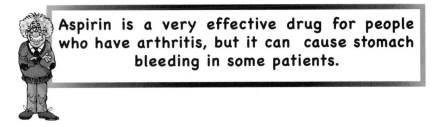

Aspirin is a very effective drug for people who have arthritis, but it can cause stomach bleeding in some patients.

Unfortunately, when the study was carefully examined, a clear deception was evident. The company that makes Celebrex only reported the results for the first six months of the **YEAR-LONG** study. When data from the entire year was analyzed, there was no ulcer-reducing benefit from Celebrex. Because people with arthritis usually take medication for years, a drug that reduced ulcers for only six months would not be very beneficial.[3]

The Cash For Clunkers "Fiasco"

The Cash for Clunkers program attracted mounting criticism when the Associated Press reported that data released under the Freedom of Information Act showed that the most common vehicle trade-ins during the government's Cash for Clunkers program were Ford or Chevrolet pickup trucks. These trucks, which accounted for more than 8,000 of the swapped vehicles, were typically traded in for new trucks whose fuel economy rating was only marginally better than the old clunkers. (1 mpg to 3 mpg improvement)

> **The Cash for Clunkers program was a 3 billion dollar U.S. Federal program that provided economic incentives to trade in older vehicles with poor mileage for new, more fuel-efficient vehicles. The goals of the program were not only to put safer, cleaner and more fuel-efficient vehicles in service, but to also provide an economic stimulus to the auto industry.**

After the Associated Press released the news that pickups were the most common swap in the Cash for Clunkers program, there were several calls for an investigation into "this wasteful government boondoggle".

If you do not selectively point out the ineffective parts of the Cash for Clunkers program and look carefully at all the data, a different story emerges. Yes, the 8,000 or so pickup trucks were the most common vehicles traded in during the Cash for Clunkers program, but they amounted to approximately 1% of the 677,081 total trade-ins processed by the government.

Later in the AP report we find the statistic that the average trade-in had a fuel economy of 15.8 miles per gallon and was traded in for a vehicle whose mileage was 24.9 mpg. This is a 58% gain in mileage! Unfortunately, the article focused on a very small part of the Cash for Clunkers program, a part that clearly was not effective. Because of this, the article left the impression that the program was marginally effective at best.

This would be similar to a situation where an archer hits a bull's-eye 99 times and misses the target once. Would it be fair to ignore the 99 great shots and do a news story focusing on the one missed shot?

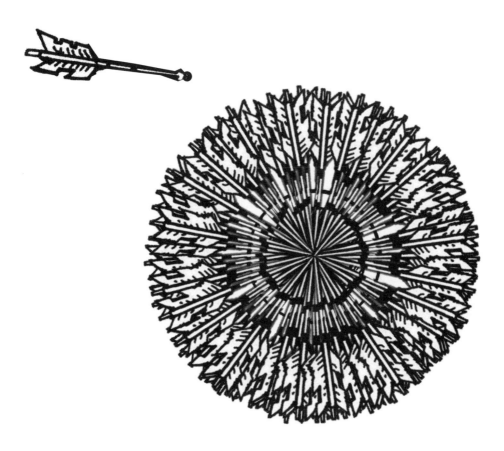

The Pollster / Think Tank Partnership

It is very rare to have an e-mail trail that documents the process of statistical manipulation done through cherry-picking data favorable to one's agenda. The following story is a clear example of this common and disturbing technique.

In 2009, the University of Wisconsin-Madison announced a partnership with the Wisconsin Policy Research Institute to conduct polling across the state. At the time, the announcement referred to the Wisconsin Policy Research Institute (WPRI) as a "non-partisan, not-for-profit think tank".

Although University of Madison school officials touted the benefits of the partnership, many others were alarmed. They said the Wisconsin Policy Research Institute was not non-partisan and was in fact the State of Wisconsin's leading right-wing think tank that had a clear free-market, limited government slant. They also pointed out that WPRI received funding from the Bradley Foundation --- a Milwaukee based foundation that supports conservative causes.

The first poll came out in October of 2009 and covered various topics including the governor's race in Wisconsin, President Obama, and school vouchers. Because the organization that paid for the polling (WPRI) favored school vouchers, they were apparently disappointed when the poll results showed there was statewide opposition to vouchers by a margin of 46.6% to 42.4%. A very interesting press release followed that did not mention the statewide polling percentages. What **was** mentioned was that a majority of people in **Milwaukee County** supported vouchers.

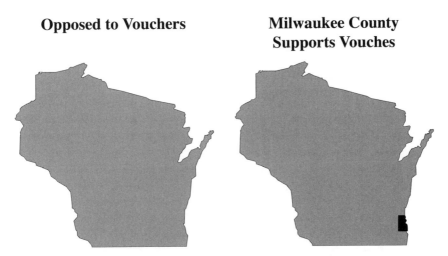

Opposed to Vouchers   **Milwaukee County Supports Vouches**

A liberal organization, One Wisconsin Now, obtained documents and e-mails between the head of WPRI and the head pollster. The documents revealed a clear attempt by WPRI to get the pollster to deliberately downplay the results of the question on vouchers. The following e-mail exchange bears this out:

Wisconsin Policy Research Institute: *"Yes, but I don't want the top line going out without it and was hoping we could incorporate this into the question. Otherwise, someone with ill intentions could use our release inappropriately."*

Toplines are the questions asked and the results. An example would be school vouchers: 46.6% opposed and 42.4% favored. In this e-mail the head of the Wisconsin Policy Research Institute is trying to get the Milwaukee County results in the topline instead of the statewide results because polling in Milwaukee County showed support for vouchers.

Head pollster: *"I understand, but i think it looks strange to put in the regional breakdown in the toplines just on that one question. we have the comparisons in the powerpoints...you can make the*

*regional document i sent a little prettier and post a separate (sic)
thing, but i think the toplines should be the toplines for the state."*

Wisconsin Policy Research Institute: *"I'm not concerned
about journalists. I'm concerned about the Scott(sic) Ross types
who would enjoy being able to portray WPRI's own data as showing lack of support for choice. I know it's a pain in the ass but I've
been burned a couple of times and I don't need to be the one hold-
ing the gas can."*

After the powerpoint presentation was altered and the statewide
results of the voucher question were replaced with the more favorable Milwaukee County results, the head of the Wisconsin Policy
Research Institute e-mailed a thank you to the head pollster:

Wisconsin Policy Research Institute: *"Thanks for the added
info. over the weekend. It helped immensely with my correspon-
dence with my board and other consumers of WPRI material."*

The polling results on the school voucher question did not fit the
pro-voucher agenda of the Wisconsin Policy Research Institute, so
the pollster was pressured to highlight the results that were "voucher
friendly". This situation highlights the problem when public polling
is paid for by an interest group. Even though the pollster who was
pressured by WPRI was known for his integrity and competence,
when an interest group pays for a poll, the accuracy and indepen-
dence of the results can easily be compromised.

Even Jay Leno Manipulates Statistics

Have you ever worried about the collective intellect of the United States after seeing a clip of "Jay Walking" from the Jay Leno Show? For those not familiar with Jay Walking, Leno walks the streets and asks basic questions of the public. Sometimes he will even go to a college graduation and ask future teachers very easy questions to make the inevitable incorrect responses even more dramatic.

The segments nearly always elicit uncomfortable laughter and disbelief that person after person could be so intellectually inept. What the audience doesn't see of course is the number of people Jay must interview in order to get the outrageous answers that he gets. (Hopefully it is a large number.)

If we saw the entire process of 50 correct answers and one wrong answer, the segment would not be very entertaining. Jay looks for people who are not bright, and he finds them if he interviews enough people.

Conversely, Jay could also ask people to explain the Pythagorean theorem or explain what a quadratic equation is. He might find one in a hundred who knew the answer, but if he shows only those clips, then the viewers will go away very impressed with the intellect of the country. Jay is "painting bull's-eyes around arrows" when he selectively picks data that furthers his agenda --- making people look not very bright.

CHAPTER FOUR
The Power of Honest Statistics

Statistics may be defined as "a body of methods for making wise decisions in the face of uncertainty."
---**W.A. Wallis**

The Statistical Warning That was Ignored --- Which Led to 40,000 Deaths

The pharmaceutical industry uses statistics not only to judge whether a new drug is effective against a disease or condition, but also to ensure that it is safe. In 1999, a large pharmaceutical company was undergoing the final stages of testing for a drug called Vioxx. Vioxx was a promising painkilling drug that had a distinct advantage over aspirin --- it did not cause gastrointestinal complications such as stomach bleeding and ulcers. Because it was "stomach safe", the potential sales for Vioxx were significant.

Prior to the final testing stage, the pharmaceutical company knew it had to choose the competition for the Vioxx study very carefully. If it chose aspirin, the proven protective effects of aspirin against cardiovascular problems risked making Vioxx appear to be an inferior drug. In other words, the test would most likely show fewer people in the aspirin group dying from heart attacks.

After considerable deliberation, the following guidelines were set up for an 8000-participant clinical trial:

• Aspirin would not be the competition.

• Anyone with a high risk of heart problems would be excluded from the study.

• Study participants would be told to not take aspirin during the clinical trial.

• People with existing gastrointestinal problems would be excluded from the trial.

• The "competition" for Vioxx in the clinical trial would be Aleve (naproxen) because it was not known to have a protective effect against cardiovascular events.

> **The results of the trial were stunning. After nine months, the group that took Vioxx had four times the number of heart attacks as the group that took Aleve. The math was very, very clear --- Vioxx almost certainly was a powerful cause of heart attacks.**

Unfortunately, because so much money was at stake, the results of the study were interpreted in a very disturbing way. Instead of acknowledging that Vioxx increased the risk of cardiac events by 400 percent, it was claimed that Aleve reduced the risk by 80%. This was an especially implausible interpretation because Aleve was not known to have a heart protective benefit, and an 80% risk reduction would make Aleve two to three times more effective than aspirin in reducing heart attacks.

This study should have prevented Vioxx from being marketed to millions of unsuspecting individuals who sought pain relief. The impending disaster could have and should have been averted by the clear message in the statistics from the Vioxx/Aleve study.

Unfortunately, Vioxx was approved by the FDA and subsequently used by millions of people. It was finally pulled from the market four years later, but not before causing a staggering number of heart attacks and deaths.

The FDA's scientists have estimated that in the four years Vioxx was on the market, it caused between 88,000 and 139,000 heart attacks of which 30–40 percent were fatal.[4]

Mathematics Versus Professional Wine Tasters

A Princeton economist named Orley Ashenfelter believed he could better assess the future quality of wine by using mathematics instead of the traditional tasting method done by "experts". (Some wines spend months in oak casks before they are bottled. Wine tasters often attempt to judge the future quality of a wine while it is still fermenting.)

Because wine is affected by weather, Orley theorized that he could predict a wine's quality by analyzing the weather data from the year its grapes were grown. After determining statistically that the best wines were produced in summers when temperatures were high and rain was low, Orley used his data to come up with an actual formula for wine quality.

> **Wine Quality = 12.145 + 0.00117 winter rainfall + 0.0614 average growing season temperature - 0.00386 harvest rainfall** [5]

Orley then unleashed the wrath of the professional wine tasters when he stated that mathematics could predict the future quality of a wine better than actually tasting the wine.

"Neanderthal way of looking at wine"
"Ludicrous and absurd"

"So absurd as to be
laughable"

"Total sham"

Ashenfelter started making predictions about the quality of wines in the 1980's. Because his formula was considerably more accurate than the traditional method of tasting wines, he slowly developed a following among wine buffs.

His big break came when he disagreed with a leading wine critic's assessment of a 1986 Bordeaux. The critic thought the wine would be excellent while Ashenfelter's formula predicted mediocrity. Ashenfelter was correct. He then predicted that a 1989 Bordeaux would be the best wine in decades before wine critics even had a chance to taste the wine. He went on to predict that 1990 would produce even better wines than 1989. Ashenfelter proved accurate both times because his formula "knew" that the weather in 1989 and 1990 would produce outstanding wines.

Even though Ashenfelter's statistical method of determining the best wines was unmercifully mocked, when his equation went head to head against the ability of the wine tasters, mathematics clearly won!

Mathematics Versus the Legal Experts

Andrew Martin and Kevin Quinn are two political scientists who devised an experiment to determine the best way to predict how Supreme Court justices would vote. In 2002, they set up a competition between a fairly crude statistical model and 83 legal experts to predict how each Supreme Court Justice would vote during the 2002 term.

The statistical model consisted of only six factors:

(1) The circuit court of origin
(2) The issue area of the case (free speech, civil rights etc.)
(3) The type of petitioner (United States, an employer, etc.)
(4) The type of respondent (United States, an employer, etc.)
(5) Was the lower court ruling liberal or conservative?
(6) Did the petitioner argue that a law or practice was unconstitutional?[6]

Did experience and intuition trump an inflexible, unthinking, mathematical model? Not even close! The statistical model accurately predicted affirm or reverse votes 75% of the time while the legal experts succeeded only 59% of the time.

How Statistics Saved 100,000 Lives in 18 Months

Don Berrick is a pediatrician and the president of the Institute for Healthcare Improvement. Two events in 1999 turned him into a passionate crusader for safer hospital care.

The first event was the publication of a report by the Institute of Medicine that documented flaws in our treatment of hospital patients --- preventable medical errors --- that caused close to 100,000 deaths each year.

The second event was the treatment his seriously ill wife experienced while she was hospitalized:

• The same questions were repeatedly asked by each new doctor.

• Drugs that were proven to be ineffective were tried more than once.

• Doctors determined that it was of the utmost importance that she quickly be treated with chemotherapy, but she ended up waiting 60 hours for the treatment to start.

• She was left alone in a basement three separate times at night.

These two events started Dr. Berwick on his quest to prevent unnecessary hospital deaths. He studied the statistics concerning patient mortality and challenged hospitals to make six distinct procedural changes that would likely prevent a significant number of those deaths. He determined that if a large number of hospitals agreed to implement his suggestions, over 100,000 lives could be saved in an 18 month period.

These are the six
changes he recommended.

(1) Elevate beds after surgery and clean the patient's mouth frequently. Studies show that if a bed is elevated after surgery and the patient's mouth is cleaned frequently, lung infections (ventilator-associated pneumonias) are dramatically reduced. (Thousands of lives could be saved with this procedure.)

(2) Stop medication errors. Institute checklist procedures to make sure the correct drugs are prescribed and administered. Make sure medicines are transferred accurately when a patient changes hospital rooms, changes hospitals, or returns home.

(3) Ensure that evidence-based heart attack treatments are used (Aspirin and beta blockers on arrival and then clot busters).

(4) Ensure that there is timely, efficient and effective bedside response when a patient needs emergency care.

(5) Implement procedures to prevent infections caused by central line catheters such as hand washing and cleaning the skin with a powerful antiseptic. This procedure alone was estimated to have the potential to prevent 25,000 deaths per year.

(6) Stop surgical-site infections --- the most common cause of complications or death after operations. Ensure the right antibiotics are used at the right time. Ensure that hand-washing rules are enforced. Clip hair at the surgery site instead of shaving, which avoids nicking the skin.

Dr. Berwick likes to compare his changes to the checklist airline pilots are required to follow. Pilots cannot choose to follow a checklist or not and he believes doctors should not have the choice either. When the procedures are done methodically and by the book, fewer mistakes occur and a significant number of lives are saved.

Berwick signed up nearly 75 percent of the hospitals in the United States. Some fully implemented the programs, while others agreed to use some of his suggestions. After 18 months, a statistical analysis showed that the new procedures saved over 120,000 lives.

Dr. Berwick had a dramatic impact on hospital safety by simply doing a statistical analysis to determine why people were dying in hospitals, and then looking for evidenced-based interventions to prevent those deaths. (For more information read *Supercrunchers* by Ian Ayres, chapter 4.)

How Statistics Helped Emergency Room Doctors Make Better Decisions

With their extensive medical training and years of experience, you might expect that doctors can tell which patients who visit emergency rooms have life-threatening conditions and which complaints are not as serious. ER doctors can usually make accurate judgments about their patients, but the application of statistics to emergency room medicine has proven extraordinarily helpful to the decision-making process.

Steven Levitt and Stephen Dubner, in their book *Super Freakonomics* (2009), tell the story of an extensive statistical analysis of over 600,000 emergency room visits by almost a quarter million patients over an eight-year period at Washington Hospital Center. (The study was led by Craig Freid.)

After the data was analyzed, the researchers were able to tell doctors which emergency room complaints result in relatively high death rates and which ones result in low death rates.

Relatively high death rate:	Relatively low death rate:
Clots	Chest pains
Fever	Dizziness
Infection	Numbness
Shortness of breath	Psychiatric complaints

Even though "chest pains" are often considered just as serious as "shortness of breath", the numbers were very clear. A year after visiting the emergency room, patients who complained of shortness of breath died at a rate 2.5 times greater than those who complained of chest pains.

When statistics are used properly, they can save lives.

Can Statistics Improve a Baseball Team?
(Baseball Scouts Versus a Computer)

Major league baseball is full of high-priced teams that are very successful year after year. The league is also filled with teams whose payrolls don't allow them the luxury of acquiring the best talent. These teams usually are perennial bottom dwellers in the league standings.

Spending large amounts of money doesn't guarantee success, though. Teams with high payrolls can be found near the bottom of each division's standings. In addition, very low team payrolls do not guarantee failure.

One of the biggest mysteries in baseball at the beginning of the 21st century was why many teams with very high payrolls failed to produce competitive teams, while a team with one of the lowest payrolls in baseball (Oakland A's), consistently ended up making the playoffs.

When the reason for the success of the Oakland A's finally became clear, it surprised even the most knowledgeable baseball fans. The Oakland A's may have had a low payroll and may not have been able to afford the best players, but they had a secret weapon. That weapon was **statistics**.

The general manager of the Oakland A's, Billy Bean, knowing he would never have the money to acquire his dream players, decided to scientifically study baseball. What did statistics say about the importance of foot speed, about home run hitters versus singles hitters, or about players who had mediocre batting averages but were adept at helping to create runs? (Which after all is the goal of every baseball team.)

To produce successful teams, Billy Bean used an innovative technique pioneered by baseball writer Bill James: data-driven decision making. The premise of this new theory was that statistics trumped observational expertise. To make his point, James pointed out that the difference between a .300 hitter and a .275 hitter is one additional hit every two weeks. Even if you watched the two players for two weeks, you would not be able to tell who was the better hitter, but a computer could.

Bill James even devised a mathematical formula to quantify a hitter's contribution to the creation of runs:

Runs Created =
(Hits + Walks) x (Total Bases)/ (At Bats + Walks)

[7]

Baseball scouts were dismissive of this detached, mathematical approach to baseball. They had experience, knowledge, and they knew how to spot talent. How could a computer compete against them in making judgments about players?

In the end, observational expertise was no match for quantitative data. The problem with baseball scouts was that they had biases that clouded their judgments. An overweight, homely, prematurely balding young man did not look like a major league ballplayer to a scout. The computer, on the other hand, didn't care if the young man looked like a circus clown or looked like George Clooney. It saw only the numbers the player produced and made judgments based solely on mathematics.

After the Oakland A's extraordinary success using statistics, more and more baseball teams started using data to guide their decision making. As they moved away from depending on observational expertise, many of these teams experienced dramatic improvements in their performance.

(For more information, read *Moneyball* by Michael Lewis.)

An Early Use of Statistics That Saved Thousands of Lives

London suffered through several outbreaks of cholera in the 19th century. Because doctors at the time did not know about bacteria and viruses, they did not understand what caused diseases such as cholera.

Dr. Snow was a London physician who treated many patients who were affected by cholera during the epidemics. He didn't understand why some people got sick and others were completely unaffected by the disease, so he plotted the cholera deaths on a map.

When Dr. Snow looked at his completed map, he discovered an unusually high number of deaths near one specific water pump on Broad Street. He wasn't sure why, but it was clear to Dr. Snow that people who used water from the Broad Street pump were at a high risk of contracting cholera. After Dr. Snow convinced local authorities to disable the Broad Street pump, the number of cholera deaths decreased dramatically.

Dr. Snow's use of statistics was one of the earliest examples of the life-saving power of statistics. The data Dr. Snow gathered about cholera deaths in London not only taught us the importance of sanitation in preventing illness, but it also showed us the importance of organizing and analyzing data.

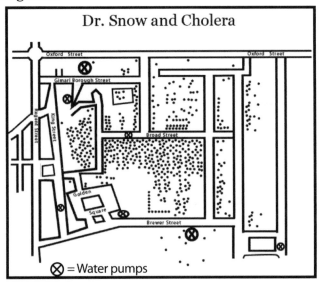

CHAPTER FIVE
Purposely Fogging Clarity

He uses statistics as a drunken man uses lampposts
- for support rather than for illumination.
--- Andrew Lang

The O.J. Simpson Trial

In 1994, O.J. Simpson went to trial for the murders of his ex-wife, Nicole Simpson, and a male companion. The prosecution had an overwhelming amount of physical evidence available to use in the trial:

  • O. J. Simpson's blood was at the crime scene.

  • Nicole Simpson's blood was found in O. J. Simpson's truck, on his socks, and in his home.

  • One of O.J. Simpson's gloves was found at the scene of the murder and the other one was found on his property.

The prosecution also had clear evidence of past physical assaults of Nicole Simpson by the defendant and they spent a substantial amount of time documenting O.J. Simpson's abuse of his former wife. Their message to the jury was clear: **A history of abuse would make O.J. Simpson a strong suspect, even if there was no other physical evidence linking him to the crime.**

When it came time to rebut the prosecution's case, the defense team used statistics to try to minimize any connection between spousal abuse and a subsequent murder. They presented the following information to the jury:

There are approximately 4 million women abused each year and in 1992, 1432 were murdered by their abusers. In other words, only 1 in 2500 abused women ended up being murdered by their abuser.

This statistic is accurate and to a jury with little understanding of how numbers can be manipulated, it was easy for them to conclude that O.J. Simpson's previous abuse of Nicole Simpson had very little to do with her murder.

Some statistics are meant to shine light on the truth or the reality of a situation while others are meant to obfuscate or fog the truth. The "1 in 2500 abusers end up killing their partner" is clearly in the fogging the truth category. There is a very important statistic the jury should have heard, but did not.

Before we get to the missing statistic (about the connection between abuse and murder), let's shift gears and look at a different debate. Patient X is a smoker and has lung cancer. Is it reasonable to say that we should suspect smoking as the cause?

The connection is not clear! There are 50 million smokers in the United States and only 200,000 new cases of lung cancer each year. Even if you smoke, there is only a 1 in 250 chance you will get lung cancer each year.

The 1 in 250 statistic is clearly meant to fog the truth because that probability is **not relevant** to the question. If the question concerned the likelihood of a smoker getting lung cancer in any given year, then 1 in 250 would be an honest and fair statistic. The question here, though, concerns a smoker who **has** lung cancer. Over 90% of lung cancers occur in smokers, which is a strong indictment of smoking as a causal element for lung cancer.

The connection is clear. Over 90% of those who get lung cancer are smokers, so we can point a finger at tobacco as a probable cause of lung cancer.

The pertinent question during the O.J. Simpson trial was not:
   "What is the probability of a battered woman being murdered by her abuser?"

That question was **not relevant** to the situation where an abused woman already was murdered. The relevant question is:
"If a battered woman is murdered, how often is the abuser the perpetrator?"

The answer to that question --- over 90% of the time!"

If an abused woman is murdered, the abuser is the guilty party over 90% of the time!

This statistic alone does not prove that O.J. Simpson murdered his ex-wife, but it does tell you that the likelihood he was involved is very high. When the physical evidence that connects him to the crime is factored in, it becomes clear that O.J. Simpson did commit the murders he was charged with and should have been convicted.

How Reye's Syndrome Caused Hundreds
of Preventable Deaths in Children

Reye's Syndrome is a deadly disease that kills or disables a significant number of those who contract it. While the cause of Reye's Syndrome is not known, there is a well established link between the taking of aspirin by children during a viral illness and the development of Reye's Syndrome.

In 1986, the FDA required that all aspirin sold in the United States carry a label that warned of the aspirin/Reye's Syndrome connection. Since that time, the number of Reye's Syndrome cases has plummeted and thousands of lives have been saved.

At first glance, this appears to be a triumph of science and medicine working together to establish the reasons why a devastating disease was occurring and then implementing an educational campaign to make sure parents were warned. The truth is not quite that simple ----- and far more troubling.

Several years before the FDA required warnings on aspirin, many scientists and people in the medical field knew that taking aspirin during a viral illness dramatically increased a child's risk of developing Reye's Syndrome. Even though the science was very clear and compelling, aspirin manufacturers delayed the FDA's response to the danger by arguing that the link between Reye's Syndrome and aspirin was not really established. They claimed that the studies that confirmed aspirin's link to Reye's Syndrome had flaws and demanded more reliable studies before the public was warned of the "possible dangers" of aspirin.

The public was not informed for over two years because of the aspirin industry's delaying tactics. The delay would have lasted longer were it not for The Public Citizen's Health Research Group, which took court action to force the Reagan Administration to enact regulations that should have been in place years earlier.

The aspirin industry purposely fogged the truth and the consequence of their unconscionable behavior was that hundreds of children died needlessly.

The Appetite Suppressant
That Caused Strokes in Young Women

Phenylpropanolamine (PPA) was a popular appetite suppressant that millions of young women used in the 1970's, 80's and 90's. Shortly after the drug came on the market, physicians started noticing that young women were suffering strokes after using PPA. Because it was so unusual for young women to suffer strokes, many physicians reported their concerns to the FDA.

In 1993, the FDA was concerned enough about the safety of PPA that it announced plans to remove it from the "safe and effective" category of drugs. The manufacturers of PPA objected and eventually the industry and the FDA reached a compromise -- there would be a thorough study of the drug and its possible connection to strokes.

Ten years later, the results of the research were definitive. Phenylpropanolamine did cause hemorrhagic strokes. The data from the research showed that women who used PPA as a diet aid had a risk of hemorrhagic stroke 16 times the normal rate.

If you thought the Phenylpropanolamine industry would admit defeat after the ten-year study clearly showed that PPA was dangerous, you would be mistaken. The manufacturers of PPA tried to discredit the research and create as much doubt as possible in order to delay PPA's removal from the market.

There are actual firms that corporations can hire that specialize in discrediting research that is damaging to their particular product. The firms really don't care if the research is done properly --- their job is to cultivate doubt.

The FDA finally had enough, though, and removed PPA from the market in November of 2000. This long overdue removal came 30 years after physicians suspected that PPA caused strokes in young women. The delaying tactics by the manufacturers of Phenylpropanolamine caused thousands of preventable strokes and devastated thousands of lives.

The Tobacco Industry - True Masters at the Art of Statistical Manipulation and Purposely Fogging Clarity

"The industry understood that the public is in no position to distinguish good science from bad. Create doubt, uncertainty, and confusion. Throw mud at the "antismoking" research under the assumption that some of it is bound to stick. And buy time, lots of time, in the bargain." [8]

Doctors observed an association between smoking and cancer as early as the 18th century, but shorter life spans and a lower incidence of smoking tended to mask the connection. By the 20th century though, tobacco's connection to lung cancer and a host of other diseases was undeniable. Study after study pointed to the serious health risks that smokers faced:

• A scientist at John Hopkins University conducted a study that showed smoking dramatically lowered one's life span (1938).

• A statistical analysis published in the British Medical Journal showed that a smoker's risk of developing lung cancer was 50 times greater than a nonsmoker (1950).

• Tar from cigarette smoke, when painted on the backs of mice, produced cancerous tumors (1952).

• 13 studies showed high cancer rates in smokers (1953).

• A large study sponsored by the American Cancer Society (1954) had to be stopped early because the results were so dramatic that research ethics required the information to be immediately released to the public.

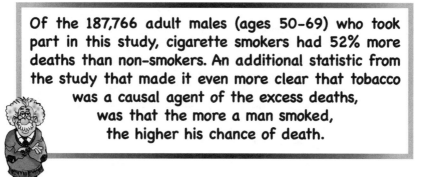

Of the 187,766 adult males (ages 50-69) who took part in this study, cigarette smokers had 52% more deaths than non-smokers. An additional statistic from the study that made it even more clear that tobacco was a causal agent of the excess deaths, was that the more a man smoked, the higher his chance of death.

The news from the American Cancer Society study that smoking was dangerous spread quickly from coast to coast and should have ended the debate over the safety of tobacco.

"Then and there, in 1954, every scientist and every executive should have said, 'Yes, more research is needed, but until we find out that these results are incorrect, let's assume that cigarettes are killers and treat them accordingly.'" [9]

The tobacco industry's response to the compelling, clear and definitive warning that statistics presented about the dangers of tobacco was to place ads in newspapers across the country.

From the tobacco industry:

**A FRANK STATEMENT
TO CIGARETTE SMOKERS**

We accept an interest in people's health as a basic responsibility, paramount to every other consideration in our business.

We believe the products we make are not injurious to health.

We always have and always will cooperate with those whose task it is to safeguard the public health.

The last statement is especially disingenuous because the tobacco industry has spent the past 75 years focusing on covering up the truth about the risks of tobacco. One of their tactics was to search for other causes of tobacco-induced diseases. The following articles, which were written for the Tobacco Institute's journal "Reports on Tobacco and Health Institute" (1961-1964), show how extensive and farfetched this search was:

• "Psychological, Familial Factors May Have Roles in Lung Cancer"

• "Lung Specialist Cites 28 Reasons for Doubting Cigarette-Cancer Link"

• "Lung Cancer Rare in Bald Men"

• "Nicotine Effect is Like Exercise" [10]

What follows is the entire text of the tobacco industry's Frank Statement to Cigarette Smokers released in 1954:

A FRANK STATEMENT TO CIGARETTE SMOKERS

Recent reports on experiments with mice have given wide publicity to a theory that cigarette smoking is in some way linked with lung cancer in human beings.

Although conducted by doctors of professional standing, these experiments are not regarded as conclusive in the field of cancer research. However, we do not believe that any serious medical research, even though its results are inconclusive should be disregarded or lightly dismissed.

At the same time, we feel it is in the public interest to call attention to the fact that eminent doctors and research scientists have publicly questioned the claimed significance of these experiments.

Distinguished authorities point out:

1. That medical research of recent years indicated many possible causes of lung cancer.

2. That there is no agreement among the authorities regarding what the cause is.

3. That there is no proof that cigarette smoking is one of the causes.

4. That statistics purporting to link cigarette smoking with the disease could apply with equal force to any one of many other aspects of modern life. Indeed the validity of the statistics themselves is questioned by numerous scientists.

We accept an interest in people's health as a basic responsibility, paramount to every other consideration in our business.

We believe the products we make are not injurious to health. We always have and always will cooperate closely with those whose task it is to safeguard the public health.

A FRANK STATEMENT TO CIGARETTE SMOKERS
(continued)

For more than 300 years tobacco has given solace, relaxation and enjoyment to mankind. At one time or another during those years critics have held it responsible for practically every disease of the human body. One by one these charges have been abandoned for lack of evidence.

Regardless of the record of the past, the fact that cigarette smoking today should even be suspected as a cause of serious disease is a matter of deep concern to us.

Many people have asked us what we are doing to meet the public's concern aroused by the recent reports. Here is the answer:

1. We are pledging aid and assistance to the research effort into all phases of tobacco use and health. This joint financial aid will of course be in addition to what is already being contributed by individual companies.

2. For this purpose we are establishing a joint industry group consisting initially of the undersigned. This group will be known as TOBACCO INDUSTRY RESEARCH COMMITTEE.

3. In charge of the research activities of the Committee will be a scientist of unimpeachable integrity and national repute. In addition there will be an Advisory Board of scientists disinterested in the cigarette industry. A group of distinguished men from medicine, science, and education will be invited to serve on this Board. These scientists will advise the Committee on its research activities.

This statement is being issued because we believe the people are entitled to know where we stand on this matter and what we intend to do about it. [11]

Asbestos: What Did Life Insurance Companies Know That the Asbestos Industry Pretended Not To?

Asbestos is a remarkable mineral that was popular among manufacturers and builders because of its resistance to heat and flame. It is so impervious to fire that clothes made from it can be dipped in fire to be cleaned.

The downside of asbestos is that it is extremely dangerous to work with and has been responsible for the deaths of millions of people worldwide.

Did the asbestos industry just not know about the dangers of asbestos, or did it react in a similar fashion as the tobacco industry? Did they react to the unfolding science about the dangers of asbestos with responsible corporate behavior, or did they try to obstruct the implementation of new regulations to protect asbestos workers? Unfortunately, the actions of the asbestos industry were similar to the tobacco industry --- they were shameful.

> **The hazards of asbestos were recognized two thousand years ago. A Roman historian named Pliny the Elder noticed that it damaged the lungs of slaves who worked with it.**

Should the asbestos industry have known that asbestos was dangerous because of the writings of Pliny the Elder? We'll give the industry a pass on that warning, but were there any other warnings that the industry should have heeded before the 1980's and 1990's? We can find that answer in insurance records from earlier in the 20th century.

> **What did the life insurance companies know about asbestos and when did they know it?**

The answer to this question will give us a pretty good idea as to when the asbestos industry should have been aware of the risks associated with exposure to the mineral and when protective measures should have been in place.

In 1918, the chief actuary of the Prudential Life Insurance Company stated that *"Asbestos workers are generally declined on account of the assumed health-injurious conditions of the industry."*

The truth was known in 1918, but the asbestos industry, instead of acknowledging the dangers, used denial, cover-up, and statistical acrobatics to delay accountability as long as possible.

> They spent millions of dollars each year not only denying the dangers of asbestos, but also attempting to keep the clear evidence of that danger out of the scientific literature and away from the public.

Insurance companies can help us determine when an activity or occupation is dangerous. If you want to know what's really dangerous, check with insurance companies -- they make their decisions based on statistics.

Is smoking dangerous? If life insurance companies charge a higher premium for smokers, you can be sure that they have solid statistics to back their reasoning.

Is piloting a private plane dangerous? If life insurance companies say it is, then it is.

Is it dangerous to be a male teenage driver? The statistics used by auto insurance companies say YES.

As a matter of fact, if you want to avoid hobbies and jobs that are dangerous, look at the questions on a life insurance form. Many life insurance companies ask if you are involved in the following activities:

• parachuting
• hang gliding
• rodeo riding
• gliding
• hot air ballooning
• flying an ultralight aircraft
• cave exploring
• mountain, rock or ice climbing
• scuba diving

Conclusion:

If statistics make your product look dangerous, there are firms and scientists who offer their services to help you. They will distort, question, and manipulate data until at the very least they will proclaim that the jury is still out as to the dangers of tobacco, asbestes, PPA, Vioxx.................

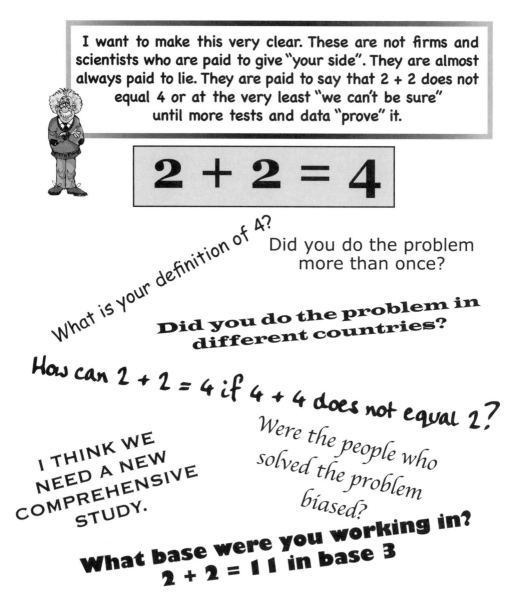

I want to make this very clear. These are not firms and scientists who are paid to give "your side". They are almost always paid to lie. They are paid to say that 2 + 2 does not equal 4 or at the very least "we can't be sure" until more tests and data "prove" it.

$$2 + 2 = 4$$

What is your definition of 4?

Did you do the problem more than once?

Did you do the problem in different countries?

How can 2 + 2 = 4 if 4 + 4 does not equal 2?

Were the people who solved the problem biased?

I THINK WE NEED A NEW COMPREHENSIVE STUDY.

What base were you working in? 2 + 2 = 11 in base 3

CHAPTER SIX
The Funding Effect

"...conflicts of interest and biases exist in virtually every field of medicine, particularly those that rely heavily on drugs or devices. It is simply no longer possible to believe much of the clinical research that is published, or to rely on the judgment of trusted physicians or authoritative medical guidelines. I take no pleasure in this conclusion, which I reached slowly and reluctantly over my two decades as an editor of The New England Journal of Medicine."

---Marcia Angell

Money Can Control Data and Restrict Public Access to Negative Results

One of the keys to proper research is the concept of "being a disinterested party". Good science requires that all data be made available, not just that which is favorable to a hypothesis. When an interested party pays scientists to do research, this basic rule can and often is violated. The following example highlights the serious consequences of this aspect of the "funding effect".

Professor Betty Dong, a pharmacologist at the University of California at San Francisco, noticed that two brand name drugs appeared to be better at treating hypothyroidism than generic drugs. After co-authoring a letter in a medical journal about the effectiveness of different drugs in the treatment of hypothyroidism, she was contacted by the manufacturer of one of the brand name drugs she had cited.

She agreed to run a clinical trial that the pharmaceutical company hoped would show that generic drugs were inferior to the manufacturer's more expensive drug in the treatment of hypothyroidism. What happened after professor Dong completed the clinical trial is an example of the abuses that can occur when actual outcomes of studies conflict with desired outcomes.

• Dong's clinical trial showed no difference between the name brand drug and the generic drugs.

• Dong sent the results of her study to the pharmaceutical company that paid for the research.

• The company claimed that her study was flawed and disputed the results.

• Dong submitted her research paper to the *Journal of the American Medical Association*.

• After undergoing peer review and slight modifications, it was accepted for publication.

• The manufacturer who paid for the study told Dong that she was not allowed to publish the results of the study without their permission.

• Dong was forced to withdraw her research paper.

Professor Dong was forced to withdraw her paper because she had signed a restrictive covenant with the corporate sponsor of her research. Restrictive covenants, which allow sponsoring corporations to control data and publication, are a major problem in research today.

Look what the International Committee for Medical Editors has to say about restrictive covenants.

"Such arrangements not only erode the fabric of intellectual inquiry that has fostered so much high quality clinical research, but also make medical journals party to potential misrepresentation, since the published manuscript may not reveal the extent to which the authors were powerless to control the content of a study that bears their name." [12]

There Can Be Serious Consequences
When Drug Companies Purposely Suppress Negative Data
(The Paxil Story)

In 1998, the medical affairs team at the pharmaceutical company that developed Paxil had a problem. Two recently completed clinical trials compared Paxil to a placebo in the treatment of pediatric depression. Unfortunately for the pharmaceutical company, there was no statistically significant difference between Paxil and the placebo.

The following quotes from a confidential memo clearly show that the medical affairs team understood what the results of the study were and how damaging they were to the company's hopes of getting Paxil approved for use by children and adolescents.

" insufficiently robust "

"failed to demonstrate a statistically significant difference from the placebo on the primary efficacy measures "

" Fail to demonstrate any separation "
(of Paxil from placebo) [13]

What was the medical affairs team's recommendation for the future of Paxil in the treatment of pediatric depression? (A drug that clinical trials showed was not much different from a sugar pill and had very real and very serious potential side effects.)

Notes from another confidential memo make it clear that the intent of the drug company was to suppress the data:

" It would be commercially unacceptable to include a statement that efficacy had not been demonstrated, as this would undermine the profile of paroxetine (Paxil)."

" There are no plans to publish data from study 377 ."

" The company should effectively manage the dissemination of these data in order to minimize any potential negative commercial impact. " [14]

This internal memo made it very clear that the makers of Paxil deliberately kept physicians and patients in the dark about the effectiveness of Paxil for children and adolescents.

When more documents were studied, it also became clear that the makers of Paxil also knew that their drug caused an increased risk of suicidal behaviors. While the company acknowledged this risk in letters to British and Canadian physicians, it withheld the information from American physicians. [15]

British authorities studied data on Paxil and discovered that children taking the drug were three times more likely to consider or attempt suicide than the placebo group.[16]

> Paxil wasn't the only drug British regulators were concerned about. "At the end of 2003, British regulators had made it clear that while the published results of randomized trials on several commonly prescribed antidepressants such as Paxil, Zoloft, Effexor, Luvox, and Remeron described these drugs as safe and effective in children, the actual data from their clinical trials showed quite the opposite." [17]

Eventually, the New York State Attorney General's office filed a lawsuit against GlaxoSmithKline (makers of Paxil) for consumer fraud. The complaint accused GlaxoSmithKline of engaging in "repeated and persistent fraud by misrepresenting, concealing and otherwise failing to disclose to physicians information about Paxil's safety and efficacy in the treatment of pediatric depression." [18]

For years, Paxil was advertised as safe and effective when evidence showed that it was neither safe nor effective. The disturbing history of Paxil shows that the funding effect is so powerful that it can lead to an industry engaging in actions that hurt our most vulnerable population --- children.

Money Can Influence and Bias Doctors

The control of data and publication is only one part of the funding effect that is slowly destroying the reliability of the clinical research published today. The other part is money --- lots of money. Drug companies spend billions and billions of dollars to influence physicians, universities, and the consumers who eventually use their drugs. Physicians are paid consultants, paid speakers, paid researchers, and recipients of gifts. In addition, drug companies also fund a majority of the continuing education classes doctors attend.

This enormous outlay of money has led to our current situation where "the pharmaceutical industry has gained enormous control over how doctors evaluate and use its own products. Its extensive ties to physicians, particularly senior faculty at prestigious medical schools, affect the results of research, the way medicine is practiced, and even the definition of what constitutes a disease." [19]

Thirty years ago, physicians who did research for pharmaceutical companies were usually intimately involved in all aspects of the research. No longer. Now drug companies control almost all parts of the research and therefore can inject their own bias to ensure that their drug appears to be effective --- whether it is or not --- and that it appears to be safer than it might actually be. (Think Vioxx from chapter 4.)

When drug companies pay doctors, design the studies, write the articles for publication, analyze the research, and decide what data the public is allowed to see and what remains hidden, we are courting disaster for our health care system.

> A professor of psychiatry, who was paid over 1.5 million dollars over a seven-year period by several drug companies, is largely responsible for the dramatic increase of children as young as two years old being treated with powerful drugs for bipolar disorder.

Antidepressant Drugs Versus Placebos ---
The Surprising Winner

Mental health professionals and patients have long assumed that there was clear and definitive scientific evidence that showed anti-depressants were effective at treating depression and that they were safe. That thinking is starting to change.

A provocative study in 1998 by Irwin Kirsch and Guy Saporstein brought to light new information that challenged the claim by phar-maceutical companies that antidepressants were safe and effective. Their study looked at 38 manufacturer-sponsored clinical studies of approximately 3000 depressed patients.

The studies showed that most of the patients who suffered from depression improved after taking antidepressants, but the control group of depressed patients who took a placebo also improved. In fact, the control group improved so much that they came very close to matching the improvement of the group that took medication.

Did sales of antidepressants plummet in response to this dramatic new information? Unfortunately, the research was either attacked or ignored and sales of antidepressants doubled over the next 10 years.

Several years later, Kirsch expanded his research by using the Freedom of Information Act to force the FDA to give him access to 47 unpublished company-sponsored studies on antidepressants. When he added these studies to his original study of 38 published clinical trials, he found that there was an even smaller benefit to antidepressants when compared to a placebo.

Kirsch felt that this research showed that those who believe that antidepressants can chemically cure depression are wrong. Kirsch also felt that this very small margin antidepressants have over placebos might even be explained by a stronger placebo response from the actual drugs. In the clinical trials, the group that received real medication experienced side effects that led them to suspect that they were not in the placebo group (80% guessed correctly). If you believe you are in the group taking the actual medication, you are more likely to believe you are being helped.

Research findings as dramatic as Kirsch's would typically inspire the scientific community to build on his research by trying to replicate or disprove Kirsch's discovery that there is very little difference between antidepressants and placebos in the treatment of depression.

Unfortunately, the response to Kirsch's research appears to have been strongly influenced by the funding effect.

• A collaboration with another scientist ended when the scientist was warned that working with Kirsch would make it very difficult to receive funding again.

• A scientist, relying on Kirsch's research, had an article published in a prestigious journal that questioned the effectiveness of antidepressants. He was subsequently warned by his department chair about the dangers of being associated with Kirsch.

Now, over 10 years later, landmark research has finally vindicated Kirsch. A study published in *The Journal of the American Medical Association* in January of 2010 showed that the response difference between antidepressants and placebos was "nonexistent to negligible in patients with mild, moderate, and even severe depression." [20]

The only patients who showed a statistically significant response to antidepressants (above the placebo response) were patients who suffered from very severe depression.

The Funding Effect Can Even Harm the
Most Vulnerable of Children ---
Those in Neonatal Intensive Care

In 1998, the FDA became concerned about the safety of a popular heartburn medicine called Propulsid. Dozens of patients died and many more suffered serious heart problems after taking the drug. What was especially alarming to the FDA was that children and infants appeared to be more vulnerable to harmful side effects from Propulsid and because of this, the FDA thought they might have to ban the drug's use in children.

After several meetings between the FDA and Johnson and Johnson, the manufacturer of Propulsid, the two parties agreed that instead of restricting the drug's use in children, new warnings would be included on the label.

Over the next two years, as deaths and injuries were increasingly tied to Propulsid, the FDA became so alarmed that it called a meeting to publicly discuss concerns about the safety and efficacy of the drug. Shortly before the meeting was to take place, Johnson and Johnson announced that it would stop selling the blockbuster drug. Four years later, Johnson and Johnson paid close to $100 million to settle lawsuits claiming that Propulsid was responsible for 300 deaths and 16,000 injuries.

Documents that were subsequently published revealed that there was a strong possibility that J&J withdrew Propulsid from the market before the FDA's public hearing because the company feared that the hearing would expose Propulsid's disturbing history. (J&J claimed that they withdrew Propulsid because physicians did not prescribe it properly.)

While preparing for the upcoming FDA hearing
which was canceled after Propulsid was withdrawn from
the market, one Johnson and Johnson executive wrote
a note that summed up Johnson and Johnson's dilemma:
"Do we want to stand in front of the world
and admit that we were never able
to prove efficacy?"

The history of Propulsid shows the serious consequences of greed, the funding effect, and the inability of the FDA to properly monitor drugs once they are approved and enter the market.

• Johnson and Johnson failed to conduct safety studies for Propulsid that were recommended not only by the FDA, but also by Johnson and Johnson's own consultants.

• Propulsid was never shown to be effective in children so it could not be advertised for use in children, but in 1998, over 1/2 million prescriptions were written for children and infants. In addition, a survey found that in 1998, 20% of babies in neonatal intensive care units were given Propulsid. [21]

> It is very disturbing that a drug with the potential for very serious side effects --- including death --- was prescribed for hundreds of thousands of children
> **WITHOUT PROOF THAT IT**
> **WAS AN EFFECTIVE TREATMENT!!!**
> Why would doctors prescribe a drug that was not effective and potentially dangerous? The answer is chilling.

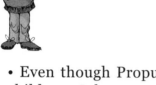

• Even though Propulsid could not be advertised for use in children, Johnson and Johnson was allowed to participate in the education of doctors, so they funded programs that encouraged the drug's use in children.

• A pediatric gastroenterologist, who supported the use of Propulsid in children, had some of his work financed by Johnson and Johnson. He also presented at a seminar sponsored by Johnson and Johnson that trained 240 doctors to talk to health professionals about Propulsid. [22]

The same gastroenterologist "edited a textbook about childhood digestive problems that recommended Propulsid." Johnson and Johnson paid for 10,000 copies of the book and distributed them to doctors. [23]

(For more information about the Propulsid tragedy, see *The New York Times*; Gardiner Harris and Eric Koli; 06/10/05.)

CHAPTER SEVEN
Poor Logic

There is an old joke about an aspiring statistician and his first trip on a plane. He was caught carrying a bomb on the plane and, when asked for an explanation, said he determined mathematically that the chance of a bomb being on a plane was one in a million. That was a little too high a probability for him, so he did some number crunching and found that the probability of two bombs on a plane was approximately one in a trillion .....so he decided to bring a bomb.

The New Chevy Volt Gets an Astounding
230 Miles Per Gallon

The Chevy Volt is a hybrid car with a very large battery pack. It plugs into household electricity to charge its batteries and will travel 40 miles on a full charge. After depleting the charge, the Volt switches to gasoline and then will travel approximately 50 miles for each gallon of gas.

The Volt is advertised (2010) as getting 230 miles per gallon, which seems extremely high considering the mileage statistics mentioned above. Let's do some experimenting and see if 230 miles per gallon is a fair claim. We'll fully charge the Volt and put a gallon of gas in its tank and determine our fuel economy as we travel.

At 40 miles, because we have used no gas, we have infinite miles per gallon.

Miles Traveled	Gas Used	Miles Per Gallon
41 miles	1/50 of a gallon	2050 mpg
50 miles	1/5 of a gallon	250 mpg
90 miles	one gallon	90 mpg

Let's fill the Volt's tank.

Miles Traveled	Gas Used	Miles Per Gallon
500 miles	9.2 gallons (40 "free" miles then 50 mpg)	54 mpg
1000 miles	19.2 gallons	52 mpg

If we drive our vehicle 51 miles, our fuel economy is the advertised 230 miles per gallon. So what is a fair mileage rating for the Chevy Volt? Certainly not 230 mpg because if we charge the battery and put one gallon of gas in the tank, the Volt will travel only 90 miles. If we fill its tank, the Volt will get 54 mpg on a 500 mile trip. Is 54 mpg fair?

Plug-in hybrids CANNOT be rated the same as conventional gas powered cars or standard hybrids. They are a new class of car! The fairest way to rate the Volt is neither complicated nor controversial:

The Chevy Volt can go 40 miles with no gas and then 50 miles per gallon after that.

Why are More and More People Contracting Diabetes ?

Parade Magazine interviewed a journalist who wrote a book about the sharp increase in diabetes in the United States.

Question: Is diabetes becoming more common because Americans are getting heavier?

Answer: "Our weight has been inching up, but diabetes is five times more common now than it was during World War II. We're certainly not five times heavier. And type 1 diabetes, an autoimmune disorder that wasn't traditionally associated with weight is increasing 2-3% each year."

The last sentence, if true, is certainly a reasonable point to bring up, but the first part of his answer is stunning!! The journalist implies that the connection between weight gain and diabetes is questionable because our weight hasn't increased fivefold like the incidence of diabetes!!

It is very odd that an author of a book about the increase in diabetes would make a statement like this. The truth of the matter, as statistics clearly show, is that being overweight 10 - 20 percent of your ideal body weight dramatically increases your chance of becoming diabetic.

Let's try the interview again:

Question: Is diabetes becoming more common because Americans are getting heavier?

Answer: YES!!!

Everyone in the United States
Will be Overweight by 2048

According to the projections of a group of scientists studying obesity in the United States, we are approaching a time when the entire country will be either overweight or obese. The study was done by researchers at the John Hopkins School For Public Health and was published in the journal Obesity. The lead author of the study stated:

"If these trends continue, more than 86 percent of adults will be overweight or obese by 2030 with approximately 96 percent of non-Hispanic black women and 91 percent of Mexican-American men affected. This would result in 1 of every 6 health care dollars spent in total direct health care costs paying for overweight and obesity-related costs."

"By 2048, all American adults would become overweight or obese, while black women will reach that state by 2034. In children, the prevalence of overweight (BMI 95th percentile, 30%) will nearly double by 2030."

Statisticians were upset over this press release, especially because it was from such a reputable source. The problem with the obesity prediction in this report is that there is an assumption of a linear increase. Using this same logic, we can predict that the average new home in the United States will be over 10,000 square feet by the year 2040 and that some future Olympian will run a mile in zero time.

I can use that same logic to make a prediction. Because the prevalence of anorexia has increased from one million in 1935 to 10 million in 2005, the entire country will suffer from anorexia sometime in the 22nd century.

So everyone will be obese and anorexic...

The intent of the "100% overweight or obese in 2048" announcement probably had the best of intentions --- to point out and publicize the dramatic increase in obesity in the United States. But the ludicrous prediction of an entirely overweight populace also undermined the credibility of the scientists who did the study.

Mystery Solved: Why Life Expectancy in Canada is Higher Than in the United States

The healthcare reform legislation of 2009-2010 produced some heated debates. One tactic of those opposed to increased government involvement in health care was to point out how inferior Canada's socialized health system was.

One of these opponents was conservative radio and television talk-show host Bill O'Reilly. After he repeatedly warned of the dangers of a Canadian style health care system, a viewer from Canada asked:

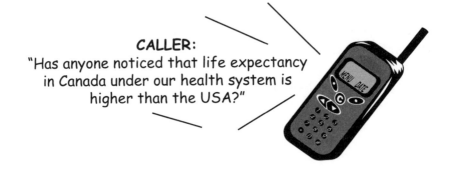

CALLER:
"Has anyone noticed that life expectancy in Canada under our health system is higher than the USA?"

Bill O'Reilly's response was very interesting.

"Well, that's to be expected Peter, because we have 10 times as many people as you do. That translates to 10 times as many accidents, crimes, down the line."

Noooooooooooo!!!!!!!

Mr. O'Reilly's statement makes no sense. The Canadian and American health care systems both have positive and negative aspects and one can make reasoned arguments as to which one is better. Bill O'Reilly's statement shows that he does not understand that life expectancy is not impacted by population size because it refers to the average age of death for a certain population.

Using Mr. O'Reilly's logic, one would assume a city of 50,000 would have a lower life expectancy than a town of 5000 because the city has 10 times the population of the town. (That translates to 10 times as many accidents, crimes, down the line.)

If the caller had argued that Canada's health care system was better because they only have 200,000 deaths each year compared to 2,000,000 in the United states, then Mr. O'Reilly's statement would have been a fairly persuasive response.

"Our country has a higher average salary compared to your country because we have more people And another thing, the only reason our obesity rate is higher than yours is because we have more people."

Advertising Hyperbole

The manufacturers of Dyson Vacuums make a very interesting claim in their advertising:

DYSON VACUUMS SPIN THE AIR UP TO 100,000 TIMES THE FORCE OF GRAVITY.

Sounds impressive unless you know that gravity is a very, very weak force. Dyson Vacuums appear to be high quality vacuums, but making a claim that they spin air 100,000 times the force of gravity is similar to an ad for a car claiming that its engine is 100,000 times more powerful than a hamster running in a wheel.

Very Strange Logic

The murder of Jon Benet Ramsey in Boulder, Colorado in 1996 attracted extensive media coverage. Years later, a man confessed to the crime and, even though his confession did not appear credible, he was arrested in Thailand and flown back to Boulder for a DNA test.

The man's DNA did not match the DNA found on the murder victim and he was released. The district attorney was subsequently subjected to intense press criticism for spending tens of thousands of taxpayer dollars to apprehend and fly the suspect back to Colorado.

During a radio interview in Texas, a relative of the murder victim was asked if it was a mistake to spend a significant amount of the taxpayer's money on what many considered a "wild goose chase". Her response:

"Sure, the DNA didn't match, but it either matched or it didn't. There was a 50-50 chance of it matching."

If that is how you view the world of probabilities, then you should be buying lottery tickets because each ticket is either a winner or a loser ----
a 50 - 50 chance of becoming a millionaire. If you think that way, though, you should be very nervous each time you go to bed, because in the morning you will be either dead or alive!

CHAPTER EIGHT
Cause Correlation Confusion

The failure to understand the difference between cause and correlation has led to an inability to differentiate what is true and what is not true in medicine, education and in many areas of our everyday lives. While correlation does suggest possible causation, it only hints at it. Without a clear causal mechanism, you cannot be sure that the relationship is anything more than a correlation.

My research led me to a discovery that I felt I must bring to the world's attention immediately. My research involved a study of where people are dying. I wanted to know the places that must be avoided in order to lessen one's chances of dying.

The results of my study are stunning! I found that 87% of people who die are in beds when they die. Because so many deaths are associated with being in a bed, I sleep on the couch now. My research shows that it is far too dangerous to sleep in a bed.

This researcher, of course, has drawn a silly conclusion from his data. He found that a high percentage of people who die are in a bed when they die, so he concluded that beds must be dangerous. His mistake, confusing cause with correlation, is very common. The researcher's thinking error is obvious here, but in many situations it is easy to unknowingly fall into the trap of assigning cause improperly.

Restoring Voting Rights to Former Prisoners Will Lower the Crime Rate

In June of 2005, Governor Vilsack of Iowa announced that he would restore voting rights to felons after their prison sentences were completed. Prisoner advocates have long asserted that the right to vote is a key ingredient of prisoner assimilation into society and they hoped that Iowa's decision to loosen voting restrictions for felons would motivate other states to follow suit.

At a news conference where he discussed the voting rights restoration, Governor Vilsack stated: "When you've paid your debt to society, you need to be reconnected and re-engaged to society." He also referenced research that showed that ex-felons who vote are less likely to return to prison.

In 2006, the state of Kentucky also debated the merits of automatically restoring voting rights to felons upon completion of their prison terms. The Voting Rights Coalition, as part of their campaign to enact felon voting rights laws in Kentucky, released research that seemed to back up their claim that voting restoration will "promote rehabilitation and reintegration into the community" and will help former prisoners avoid further contact with the criminal justice system.

The Voting Rights Coalition quoted from an article in the Columbia Human Law Review (2005) titled "Voting and Subsequent Crime and Arrest: Evidence from a Community Sample". The authors of the study, Christopher Uggen and Jeff Manza, found differences between voters and nonvoters in rates of arrest, incarceration and self-reported criminal behavior.

They also found that from 1997 to 2000, 27% of nonvoters with a prior arrest were re-arrested while only 12% of voters with a prior arrest were re-arrested.

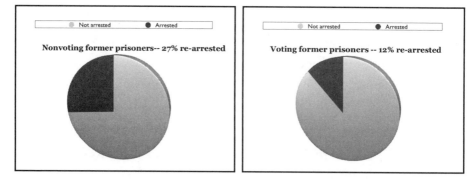

There are many good reasons to restore voting rights to prisoners who have re-entered society. Governor Vilsack probably said it best when he talked about the importance of reconnecting and re-engaging into society. The flaw in the argument of many of the voting rights advocates is that they imply that restoring voting rights will lead to less recidivism.

The fact that you can take a cohort of former prisoners who did not vote and find 27% were re-arrested while only 12% of the former prisoner voting cohort were re-arrested does not allow you to assign a causal effect to voting.

We cannot say that voting (A) causes less criminal activity (B). The most likely explanation for the lowered recidivism is that former prisoners who have decided they want to be law abiding citizens are more likely to vote than those who have not decided to assimilate.

In other words, voting is probably a "symptom" of a former prisoner who has decided to be a good citizen. The desire to stop criminal behavior drives voting; voting does not drive good behavior.

What if we studied former prisoners who play golf and compared their recidivism rate to those who don't play golf. Let's say we found that 27% of non-golfers were re-arrested and only 12% of golfers were re-arrested? Could we assign a causal effect to golf?

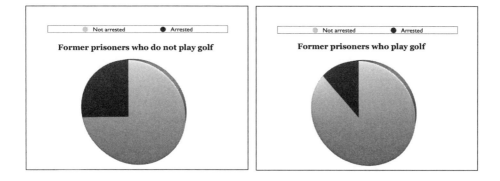

Of course golf (A) probably does not lead to less criminal activity (B). Former prisoners who have successfully re-entered society are more likely to be gainfully employed and therefore more likely to play golf. In the example, playing golf is a "symptom" of a former prisoner who has reintegrated into society.

The merits of restoring voting rights to former prisoners is not the issue we have been discussing. The question we are dealing with is whether returning voting rights to former prisoners will make it less likely that they will re-offend. Unfortunately, there is currently no research showing that voting rights restoration will help prevent recidivism.

Cause Correlation Confusion Can Cost Lives

Springfield DAILY NEWS 2009

"Medical authorities in Canada are rethinking seasonal flu shots because of new data that appears to connect seasonal flu shots with an increased risk of contracting swine flu."

In the fall of 2009, as countries prepared for the upcoming flu season, some troubling data came out of Canada that appeared to connect seasonal flu shots with an increased risk of contracting swine flu. This troubling connection was first noticed in the British Columbia Center for Disease Control during a study of the effectiveness of the previous year's flu vaccine.

During the study, it became apparent that there was a higher rate of swine flu in those who had been vaccinated for seasonal flu when compared to those who had not been vaccinated. This started an investigation and eventually researchers discovered that this connection was present in data bases across Canada.

When the seasonal flu shot/swine flu connection became apparent to medical authorities in Canada, (and before any publication of the data in a peer reviewed journal) medical decisions started to be made based on the perceived connection.

Most of the provinces and territories in Canada decided to give flu vaccinations to people 65 and older because getting the seasonal flu for this population can be life-threatening. (Those 65 and older were also at lower risk for swine flu.) The rest of the population was told that they would not be able to get a seasonal flu shot in the fall. The seasonal flu shot would become available to those who wanted it in January of 2010 if it was determined that it was safe or if the seasonal flu shot/swine flu connection was disproved.

Before we look at how Canadian authorities possibly confused cause and correlation, let's look at several problems with their response to the potential connection between seasonal flu shots and a higher incidence of swine flu.

(1) "Whatever the actual content of the British Columbian study, word of it seems to have spread with all the scrutiny and peer review of a game of telephone."[24]

(2) People making medical decisions were not only relying on an unpublished study --- many of them had not even seen the data.

(3) No biological reason was ever proposed that explained how it was possible for a seasonal flu vaccination to increase a person's risk of contracting swine flu.

Is it likely that seasonal flu shots (A) → cause → swine flu (B)

or is it a correlation? We first have to ask ourselves the following: Who gets seasonal flu shots and why?

• The elderly and those with compromised immune systems

• People who are at high risk for getting seasonal flu. (Health care workers, teachers, those who travel, etc.)

Those who get a seasonal flu shot because they are at high risk for seasonal flu are also at high risk for contracting swine flu --- not because of the flu shot --- **but because of the very reasons they decided to get a seasonal flu shot in the first place!**

Those who do not get a flu shot because they are at low risk for contracting seasonal flu are also at low risk for getting swine flu for the very same reasons they have a low risk for seasonal flu. (Not because they decided to forgo a flu shot.)

> The population that chooses to get flu shots is at high risk for any flu virus. It is not surprising that they would have a higher incidence of swine flu than an unvaccinated population. Also, those who get flu shots are probably more connected to the health care system and would be more likely to seek medical care if they contracted swine flu. Their cases of swine flu would then be recorded in the data base while others with little contact with the medical system would be less likely to seek medical attention if they came down with swine flu.

After a nationwide study, my data clearly shows
that those who wear raincoats are more likely to get
hit by hail than those who do not wear raincoats.
Something about raincoats seems to attract hail!

Clearly the raincoats do not make you more susceptible to being hit by hail. The people with raincoats have "preselected" themselves as to needing protection from rain. If they need protection from rain, they are also the ones most likely to be outside when there is inclement weather and are more at risk when there is hail.

The Key to Academic Success is Getting Parents to Attend Parent Teacher Conferences

Research has shown that the academic achievement level of students whose parents attend parent-teacher conferences is higher than the academic achievement of students whose parents do not attend the conferences.

When parents attend parent-teacher conferences, achievement is higher so we must do everything we can to get our parents to attend these conferences!

PRINCIPAL

In response to this "dramatic" research, a number of schools put in place rewards programs to entice parents to attend the school conferences. The rewards, which included money and gift certificates, did improve attendance at the parent-teacher conferences, but is it likely that student achievement improved?

Probably not. The most likely reason that students whose parents attended parent-teacher conferences had higher achievement is that education was probably a higher priority in their homes than it was in the homes of students whose parents did not attend the conferences.

Higher achievement (A) was probably not caused by the actual attendance at the conference (B). The level of importance given to education in each student's home (C) was almost certainly the important variable in student achievement.

Music and Higher Achievement

Several experiments have shown that there may be a connection between musical training and intellectual development. One recent experiment divided preschool children into three groups. The first group received keyboard instruction and singing lessons; the second group received computer lessons; and the third group received no lessons.

The results of this experiment were very interesting. The group that received the keyboard training and singing lessons scored 34% higher on tests that measured spatial-temporal ability. This connection between music training and intellectual abilities makes sense because music has clear mathematical relationships and proportions.

Another often mentioned piece of evidence for the connection between music lessons and higher intellectual development is the fact that children who have had music lessons, score significantly higher on SAT tests when compared to children who have not had music lessons.

At first glance, these test score results appear to prove that music lessons lead to better academic achievement, but we have to be careful not to confuse cause and correlation.

We need to ask ourselves whether there is something special about families that provide music lessons for their children. Because music lessons require an enormous investment in time and money, it is clear that there is something special about these families. It may turn out that music lessons do lead to higher achievement, but we can't say they do based only on the fact that children who take music lessons do better in school.

Conclusion:

Because correlation does suggest possible causation, we must ask ourselves several questions to determine whether there is in fact a causal mechanism in the relationship between A and B. Take the example of tobacco. When scientists tried to determine whether tobacco causes lung cancer, they asked the following questions:

1) Is there a correlation?

2) Does the risk of lung cancer go up when the amount of smoking increases?

3) Is the connection between lung cancer and smoking seen in other countries and in different groups of people?

4) Are different income groups affected in a similar manner?

5) Do animal studies show a connection?

6) Is it biologically plausible that smoking causes lung cancer?

To emphasize the lessons of this chapter consider the following statements and try to determine why they are true:

(1) Children with bigger feet spell better.

(2) Countries that add fluoride to water have higher cancer rates than countries that do not add fluoride.

(3) Children are more successful at school when mothers stay home and do not work outside the home.

(4) Studies show that married people are happier than those who are not married.

(5) Those who buy flood insurance are more likely to have their house destroyed by a flood.

(6) Leaving a night-light on in a child's room is connected to nearsightedness.

Answers

(1) Children with bigger feet spell better.

They are older.

(2) Countries that add fluoride to water have higher cancer rates than countries that do not add fluoride.

Wealthy countries, which usually have longer life expectancies, add fluoride to water. Their cancer rate is higher because cancer is typically a disease of old age.

(3) Children are more successful at school when mothers stay home and do not work outside the home.

Higher income levels allow the mother to stay home. Higher income is connected to higher academic achievement.

(4) Studies show that married people are happier than those who are not married.

Unsure of the answer here. Maybe unhappy people are less likely to be married.

(5) Those who buy flood insurance are more likely to have their house destroyed by a flood.

They buy flood insurance because they are in areas prone to flooding.

(6) Leaving a night-light on in a child's room is connected to nearsightedness.

A well publicized study did show a fairly strong correlation between leaving a night-light on in a child's room and the development of nearsightedness. Additional studies could not reproduce those results, but did find a strong correlation between nearsighted parents and nearsighted children. Also, nearsighted parents were more likely to turn a night-light on in a child's room.

CHAPTER NINE
Show Me ALL the Data

Do not put your faith in what statistics say until you have carefully considered what they do not say.

--- **William W. Watt**

Electing President Obama Ruined the Stock Market

The Dow Jones Industrial Average is an index showing trading averages for a number of large US-based companies. The Dow is often used as an indicator of our economy's strength, as higher trading activity tends to indicate a smoother running economy. Because the Dow consists of only numbers (which are by themselves unbiased), one would think that its reports would remain straightforward and transparent. Deception, however, is relentless when the media has an agenda to maintain.

Any graph that has enough data points can be cropped to show whatever trend its manipulators want to argue. For example, on two separate occasions, MSNBC displayed two different graphics of the Dow Jones Industrial Average, arguing that the stock market has been steadily decreasing ever since Barack Obama was elected President of the United States.

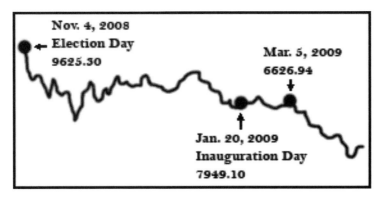

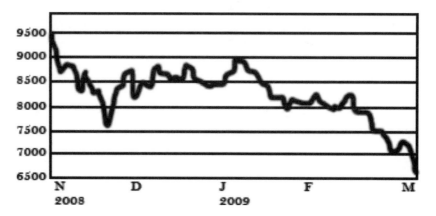

Despite obvious flaws in their design, such as the first graph having no x or y-axis, these graphs by themselves suggest that the American voters made a huge mistake by choosing Obama, as our stock market has dropped by around 30% since his election in November. On MSNBC, Alex Witt showed these Dow Jones graphs, stating that we "have a graphic showing about how Wall Street has been just staggering. So, how long do you think good intent and this relative honeymoon period will last if the numbers and the graphs keep continuing the way they look right now on our screen - downward?"

Witt's remarks would be accurate if the Dow Jones had remained steady in the months and years before President Obama was elected, or if the period of time in her graphs indicated a significant shift from earlier records. Fortunately, we do have access to an extended graph of Dow records.

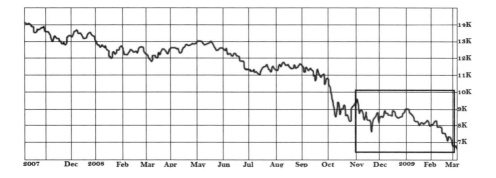

MSNBC failed to show this extended graph, which sheds light on why the stock market has dropped since Obama was elected. Looking at the full picture, we can see that the Dow Jones has been steadily declining since at least late 2007, long before Barack Obama was in the White House. The rectangle in the bottom right of the graph indicates which portion Alex Witt cropped out to make her point. When you take hundreds of data points and choose the few that support your argument, you manipulate the numbers by suggesting that they are the only data points relevant to the situation.

For example, if a member of a liberal media source wanted to show how George W. Bush has hurt the stock market, he could use the same data that MSNBC had access to, but crop a different portion of the graph. The following factually accurate graph suggests that Bush's support of the 2008 bailout plan, as well as his meeting with French President Nicolas Sarkozy, had devastating effects on the US economy.

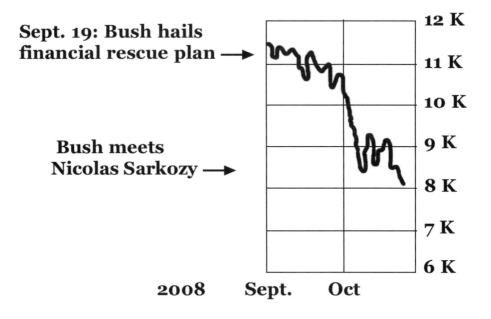

Dow Jones Industrial Average

**Sept. 19: Bush hails
financial rescue plan** ➝

**Bush meets
Nicolas Sarkozy** ➝

12 K

11 K

10 K

9 K

8 K

7 K

6 K

2008   Sept.   Oct

Advertisers and the Selective Use of Data

Advertisers can cleverly use selected sharing of information to influence potential customers. If you live near a community that has casinos, you might have seen billboards like the one shown below.

Or you might have seen articles in your local paper such as:

GOLDEN VIEW

"Local man wins $300,000 lottery. He plans to spend the money on a car, a vacation to Hawaii........."

What you don't see in the billboards and newspaper articles is all the heartache and tragedy involved in gambling. The next time you see a billboard or newspaper article extolling the gambling success of whomever, visualize several companion billboards or newspaper articles so you get a more balanced view of gambling. Think about ALL the data!

THE HERALD

.....Anna's losses at the dog track have ruined her financially, so she is thinking about embezzling money from her employer next year.... "I know the dogs, now I just need a little luck."............

Which Hotel Would You Choose?

If you saw the following advertisement, what other information would you need before you could make a judgment about how guests felt about each hotel?

Ad in paper:
"There are only two hotels in Deceptionville. Our hotel has increased its customer satisfaction rating by 20 percentage points this year while our competition saw its satisfaction rating drop three years in a row! If you want to spend the night at a hotel people are increasingly giving higher marks to, stay with us."

This ad does not give enough information to judge which hotel has the better quality. Sure, the first hotel's satisfaction rating has increased by 20 percentage points, but it could have started at 5% and gone to 25% while its competitor's ranking started at 99% and went down to 96% over a three-year period, a statistically insignificant drop.

The hotel that advertises its increased satisfaction ranking would certainly not show you these graphs:

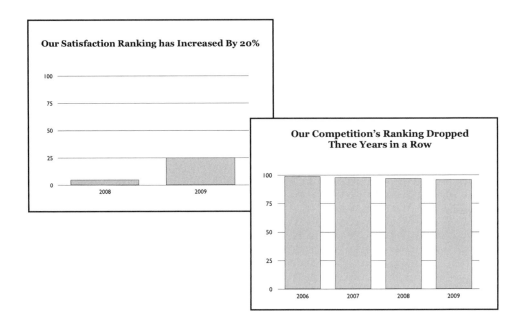

The first hotel might show you this graph:

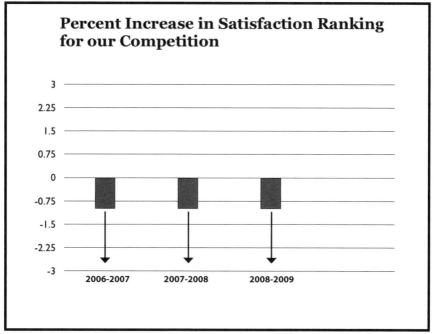

Should I Buy Gold?

In 2010, advertisements for gold started appearing in the media at unprecedented levels.

Over the last 5 years, gold has outperformed the S&P index by over 100%!

Gold is predicted to reach levels of $2000 - $3000 per ounce!

Gold is the only safe protection against inflation!

So, is gold a good investment? No one really knows for sure. We could be in the midst of a "gold bubble" similar to the technology bubble and the housing bubble, or gold really could double or triple in value in the next few years.

The investor who is trying to make a decision about the wisdom of investing in gold should always look at all the data and information about gold, not the selective pieces advertisers present. Look at the following bar graph that shows how dramatically the price of gold has increased. Even though this graph is accurate and appears to reinforce the idea that gold is a good investment, the graph does not tell the whole story.

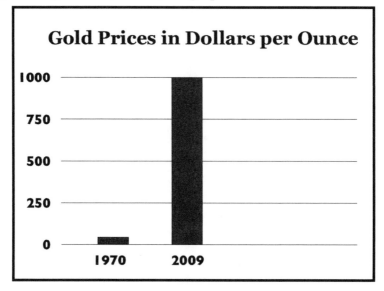

The following graph might temper one's enthusiasm for gold because it shows data that the first graph conveniently omitted. Even though gold has done well over the past few years, it is still well below its value of 30 years ago in inflation adjusted dollars.

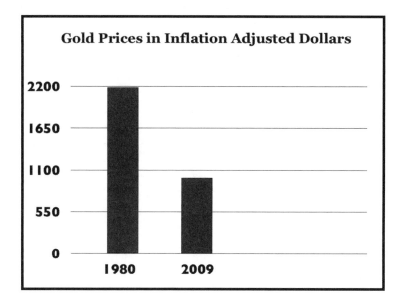

The two bar graphs show how easy it is to make gold appear to be a good investment or a bad investment by selectively picking data that reinforces a desired point of view.

Another interesting piece of information about the price of gold is that even though those selling the precious metal are accurate when they say that gold has outperformed the S&P index by over 100% in the last five years, in the past 30 years, the S&P index has outperformed gold by a wide margin. (Gold rose from $800 per ounce to $1200 per ounce while the S&P index went from $115 to $1127.)

Note: Both prices are not adjusted for inflation.

A Fair Graph With the Potential for Manipulation

The bar graph below is a good example of data presented fairly.
(Data based on a Dubuque Telegraph Herald article 6/21/2007)

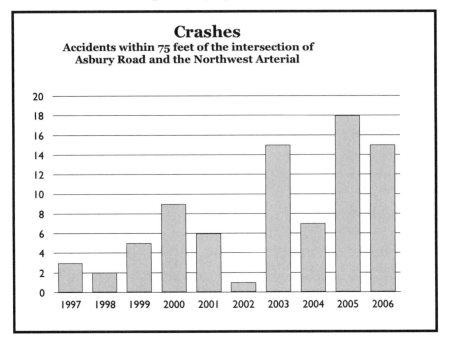

Notice how easily an individual could selectively use the data shown to further an agenda.

The intersection is getting more and more dangerous

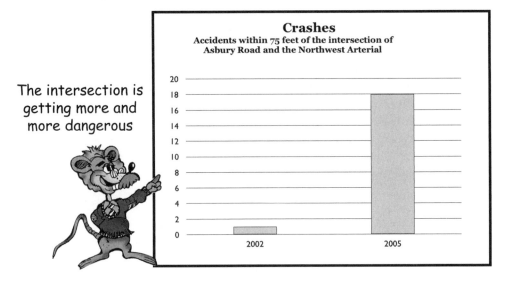

Here is another graph that shows the intersection is getting more dangerous:

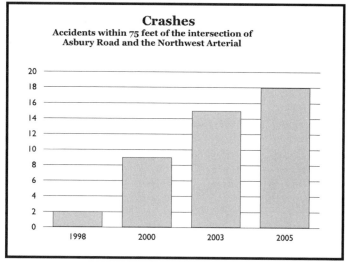

Crashes
Accidents within 75 feet of the intersection of
Asbury Road and the Northwest Arterial

If you wanted to give the impression that the intersection is not getting more dangerous, you could present the following graph:

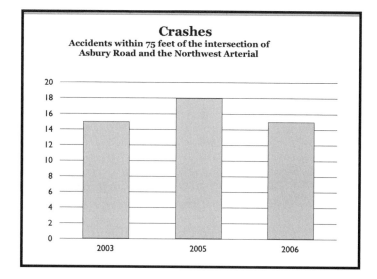

Crashes
Accidents within 75 feet of the intersection of
Asbury Road and the Northwest Arterial

CHAPTER TEN
Confirmation Bias
(Seeing What You Want to See)

"The moment a person forms a theory, his imagination sees in every object only the traits which favor that theory."
- **Thomas Jefferson**

The human brain is not only designed to find a cause for each event, but also to simplify this cause as much as possible. The evolutionary process makes it much more of an advantage to see a pattern where none exists, than to not see a pattern where one does in fact exist. In other words, there is survival value in seeing patterns, even if you are wrong --- as long as you are occasionally right.

This ability to find patterns, has led to an increased knowledge of our world and to many dramatic discoveries that have helped civilization progress. Unfortunately, the ability to seek and find patterns has also produced a significant amount of nonsense.

The Constellation Effect

Our ancient ancestors must have been very frightened by our night sky. Without the scientific understanding we have today, the peculiar balls of light in the night sky would have seemed alien and possibly frightening. To cope with the uncertainty surrounding these mystery lights, our ancestors found patterns in stars and connected them to familiar objects. These groups of stars that appeared to be the same as familiar objects are called constellations.

Records of constellations, which can be in the shape of objects such as coins, pottery and people, go back as far as 6000 years ago. An interesting thinking trap known as astrology revolves around using constellations to predict human behavior. Astrology teaches that each person's destiny is based on what constellations his or her birthday is associated with.

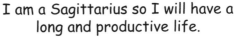

I am a Sagittarius so I will have a long and productive life.

Even though astrology seems silly and has no scientific basis, people who believe in astrology are usually neither delusional nor trying purposely to deceive anyone. They have simply fallen into a thinking trap that is a consequence of the pattern-seeking structure of the human brain. The following story is a very interesting example of this pattern-seeking structure:

The Cat That Can Predict Who Will Die

Oscar the cat has curled up in the bed of a nursing home patient named Stan. The attendants at the home now know it is time to call Stan's family because Oscar has the remarkable ability of predicting which patients will soon die.

People who witness patients dying after the visit of a cat or dog use these strange occurrences to reinforce their beliefs that a dog or cat can predict who will live and who will die. Unfortunately, the staff at Stan's nursing home probably chose to ignore all the times that Oscar the cat was too busy eating, sleeping, or licking himself to be present at the bedside of most patients as they neared death.

The theme of dogs or cats predicting who will die is a very common story and has been around for at least the last 200 years. Could animals have an extra sense that gives them the ability to predict when people are dying? There are many people who claim this is true, but it has never been scientifically demonstrated.

So we are left with several choices to explain why Oscar the cat appears to be able to predict who will die.

Super Predictor Cat

One possibility is that the cat can actually predict who will die. Oscar the cat may actually prefer sleeping in the bed of motionless patients and therefore has been around several patients who soon die.

Boring Normal Cat

I visit everyone! You just
remember the patients who
die after I visit.

Nefarious Cat

Another possibility is that the cat has some direct involvement in the patients' deaths.

Which one of these three choices is the most likely explanation for Oscar the cat appearing to have the amazing ability to predict who will live or die? Of course the most likely explanation for Oscar's apparent ability is the second option. Our tendency to look for patterns can lead us to jump to the conclusion that Oscar has a special ability when in fact he probably does not.

The way people interpret the patterns they see can be very biased. When people expect to see certain patterns, it leads to an inclination to interpret events in a certain way that confirms this biased thinking. In these circumstances, evidence that confirms the expected pattern receives a great deal of importance, while opposing evidence is ignored.

Because people think that Oscar the cat can predict who will die, they start noticing all the times that he was around before a death and ignore the times when he sat on the beds of people who did not die.

Even though people often see patterns where none exist, the ability to see patterns is a very important ability to possess. If eating a bright red berry in the forest is followed by a violent illness, the conclusion that bright red berries should be avoided has great survival value. Noticing patterns can help societies in addition to individuals. In 1796, Edward Jenner noticed that those who were infected with cowpox appeared to get only a mild case of the more deadly smallpox. This was a possible pattern. After some experimentation, it was shown that Jenner's hypothesis was accurate.

Where is the Denominator?

Let's go back to the case of Oscar the cat and take a look at how to decide scientifically whether Oscar the cat really has special powers. To do this, we will need to know a piece of information that is almost always omitted when stories like "Oscar the cat" appear in the media ... the number of events, or the denominator! The meaning and importance of this missing piece of information becomes more clear if we look at another example. Suppose a group of people saw you breaking a tree branch by simply staring at it.

People would certainly be in awe of your seemingly amazing power to break a branch as long as you did not tell them how long you had been staring at the branch WITHOUT breaking it.

I know that it appears that he has great powers, but what he isn't telling you is that he has been staring at that branch for 18 years until a gust of wind happened to come along and break the branch.

In fancy statistics talk, if you leave out the denominator and only report the numerator, you can make something appear to be true when in fact it is not true. What this says is that it is not possible to make a judgment about something when the rare is reported and how long or how many trials it took is not reported.

Would you be impressed if I rolled seven dice and ended up with all sixes?

Would you still be impressed if I told you it took 36,782 rolls?

I heard the phone ring and thought I was going to get bad news.
Sure enough, it was my grandmother telling me that
she had the flu. I must be psychic!

Before you decide to be impressed with his psychic abilities, you need to know how often he thinks a ringing phone brings bad news. It could be that he thinks something bad happens every time the phone rings.

When the media reported the dramatic story of Oscar the cat, the total number of visits that Oscar made to nursing home patients was not stated. If you are amazed that Oscar was present shortly before ten people died, you have fallen prey to numerator-based statistics. You are ignoring the fact that he probably visited each patient numerous times before his or her actual death.

We have the numerator, now what is the denominator?

10 deaths!!!

Don't forget a vital statistic when trying to decide whether Oscar is a "miracle cat"---**The denominator!!!** Anecdotal evidence is insufficient when trying to prove extraordinary events. When we have preconceived notions about certain things, we tend to latch on to the occurrences that reinforce our beliefs and ignore or downplay those that do not. As a result of their "bias", the nursing staff probably saw an imaginary pattern between Oscar and the dying patients and attributed this to a feline sixth sense.

Confirmation bias is so powerful that you can feel the effects at a stop-light. Try predicting when the light will turn green. You will be wrong far more often than you are right, but notice the thrill when you predict correctly when the light will change to green.

The times I am wrong feel insignificant compared to the thrill of being correct.

People frequently make predictions and shrug them off when they are wrong, but take a lot of credit for their "powers" when they happen to be right. This tendency to downplay incorrect predictions and to pay more attention to correct predictions is a very important surviv-al mechanism. Because our history as a species began with foraging, those with the ability to see patterns in our environment had a better chance of survival compared to those who did not see patterns.

In order to survive, individuals had to be aware that a rustling bush might mean a tiger is lurking nearby. One event of a rustling bush and then a tiger attacking would forever embed in the individual the danger of a rustling bush.

Even though a hundred future rustling bushes produced no tiger, the significant event of the rustling bush/tiger con-nection would be prominent in the brain while the "false alarm rustling bushes" would receive little importance.

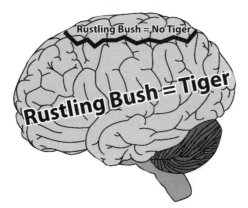

If people lingered on the insignificant occurrences (rustling bushes with no tiger) they would leave themselves open to more dangers. As a result, we now have brains that often see patterns where there are none. Here are some more examples of miraculous events that have circulated in the modern media that have their roots in the pattern-seeking architecture of the human brain.

Music Embedded in Paintings

In the midst of the popular novel "The DaVinci Code", there were many books and History Channel specials that claimed to know the "real" story. Around this time, an interesting story appeared in the media about an Italian musician who supposedly found a musical score hidden in Leonardo DaVinci's painting of the "Last Supper". Apparently he found the musical score when he took elements of the painting that have symbolic value in Christian theology and placed musical notes at these locations.

If this appears to be "forced", that is because it is. The musician could have just as easily used the feet of the apostles. All that is left now to complete the discovery of music embedded in the "Last Supper" painting is to superimpose a musical staff on the painting. But where should the musical staff be placed?

Of course the musical staff was placed where the most pleasant music would occur. When these notes are played, the result is a very pleasant musical tune.

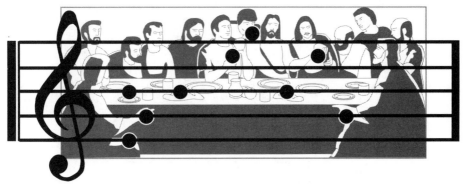

note: This is a rough drawing of the actual painting, these notes will probably not produce pleasant music.

Just what is the probability that a hidden musical score is really hidden within the DaVinci painting of the "Last Supper"? Let's look at another famous painting and see if it can clarify our thinking. Here is a cartoon depiction of one of the nine famous paintings of "Dogs Playing Poker" by C.M. Coolidge.

Could there possibly be a hidden musical score in this painting? If we can manipulate notes and superimpose a musical staff on this painting to make a pleasant tune, then we can probably do the same thing to any painting.

Using the same logic as the Italian musician who claimed to find music embedded in the "Last Supper", we need to find symbols to justify our note placement. The eyes of the dogs would seem appropriate because they are thought to be the "windows to the soul", and music is healthy for the soul. We will put musical notes over every visible eye in the painting. If you look closely at the painting, you will see that it has several glasses of liquid. The word "glass" has five letters and so does "music", so obviously the painter intended the glasses to be another symbol for musical notes!!

note: This is a cartoon depiction of the actual painting, these notes will probably not produce pleasant music.

Now we'll overlay a musical staff wherever we want and it is now obvious that the artist inserted a hidden musical score in this iconic painting! These notes make some very pleasant music.

This is all silly of course, but it makes a very important point. It shows how easy it is to find patterns when you want to find them. Our next story that deals with confirmation bias revolves around a well known book, movie, and philosophy.

The Secret

The main idea behind "The Secret" is the so-called "Law of Attraction" which states that if you focus on what you want in life, the universe will eventually rearrange itself to conform to your wishes.

Are you tired of getting bills in the mail? Well, the reason you are getting bills is because you are constantly thinking and worrying about them. This negative thinking pattern is drawing bills to you.

The way to get out of debt, according to the authors of "The Secret", is to stop thinking about your bills. Some further money tips from "The Secret":

• Do not speak of the lack of money for a single second.

• Affirm to yourself every day that you have an abundance of money, and then it will come to you effortlessly.

• Wealth is a mindset. Money is literally attracted to you or repelled from you. It's all about how you think.

Health problems? "The Secret" has answers for health issues too.

• Do not speak of your illness or disease with others.

• See yourself as only well.

• As you love completely and feel the joy within you, disease cannot exist.

You may be wondering why millions of people readily accept many of the ideas put forward by "The Secret". When this phenomenon is looked at closely, it is evident that the popularity of "The Secret" is explained by its reliance on confirmation bias and numerator based statistics.

$$\frac{\textbf{Number of times something occurs}}{\textbf{Number of opportunities for something to occur}}$$

The official Secret website is full of stories of people who wrote in with miraculous accounts of receiving various amounts of money. One report tells of a person whose former employer owed her

$5000. After reading "The Secret", she wrote herself a check for $5000. (Writing yourself checks and constantly looking at them is one tip from the book.) After writing herself the check, she visualized going to the mailbox and finding the employer's real check. Two months later, the check was in the mailbox! Of course she looked upon this happy outcome as proof that the universe "compelled" her employer to pay her.

In this case, each time the women failed to find a check in her mailbox she essentially ignored the event. When the money finally arrived, she declared it a miracle! In this example, the check came in two months, but it wouldn't have mattered if it took 2 years or 20 years because an essential part of the philosophy of "The Secret" is that the time frame is infinite.

If you gave a monkey a computer, a keyboard, and infinite time, it will eventually type out the entire United States Constitution.

"It was a dark and stormy night......

If you want a chicken sandwich to be brought to your room, you can sit and visualize the sandwich for days, months or even years until a friend brings one by. When it eventually comes, you can claim you compelled the universe to bring it to you.

Another point to remember about the $5000 miracle check is that there were undoubtedly thousands of people who also visualized coming into money and received nothing. Again, if you ignore the denominator, you cannot make a good judgment about how miraculous something is.

In addition to making sure we take into account all the people who focus on what they wanted and did not receive it, we must also remember that there are probably countless people who wanted something and received it without thinking in the way "The Secret" requires.

Another interesting example of what happens when you are given a long time span is the number of heads flipped in a row when you flip a coin numerous times. In fact, if you flip a coin $10^{1,000,007}$ times, you can expect several sequences of 1,000,000 heads in a row!!

I know this is hard to believe, but $10^{1,000,007}$ is a lot of flips. More flips than atoms in the universe!! In fact it is estimated that there are 10^{87} atoms in the entire universe, so $10^{1,000,007}$ is an awful lot of coin flips!!

The Dark Side of Confirmation Bias

While examples such as Oscar the cat and "The Secret" are somewhat humorous and shed some light on the natural biases we all have, confirmation bias can also have some very serious consequences. The following story details a study of how mental health professionals fell prey to confirmation bias and therefore were unable to accurately judge the mental health of several of their patients.

Because the National Institute of Mental Health estimates that almost 25% of adults suffer from some kind of mental disorder in any given year, it is imperative that mental health professionals accurately assess and treat their patients.

When we have plumbing problems in our homes or mechanical problems with our cars, we hope that the professionals we seek out to fix these problems accurately diagnose the problem and then undertake the needed repairs. Likewise, when we are sick, we also trust that doctors will accurately diagnose and treat our illnesses.

A plumber or a physician usually has a much less complicated task than does a mental health professional. Diagnosing a mental abnormality is much more difficult than diagnosing a broken bone because much of what we see as abnormal is dependent on where we live and the time and place the behavior occurs. For example, in some cultures eating cats, dogs or horses is considered normal. If this happened in the United States, it might suggest a mental health issue that needed to be addressed.

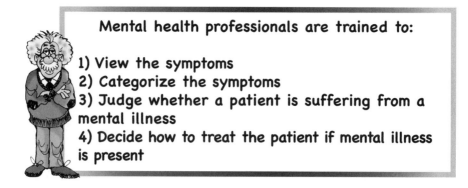

Mental health professionals are trained to:

1) View the symptoms
2) Categorize the symptoms
3) Judge whether a patient is suffering from a mental illness
4) Decide how to treat the patient if mental illness is present

A famous study by David Rosenhan makes it clear that mental health professionals often alter their judgments to fit what they expect to see. In other words, they fall victim to confirmation bias. Rosenhan's experiment took place at 12 different hospitals over varying time periods. He admitted eight people who were not suffering from any mental health problems. To gain admission to the hospitals, they were instructed to tell the hospital personnel upon admission that they were hearing voices saying the words "empty", "hollow" and "thud". They reported no other symptoms.

Once the "pseudo patients" were admitted to the hospital, they immediately stopped all symptoms and exhibited normal behavior. They told the hospital staff that they felt fine and no longer heard voices. You would hope that it would quickly become clear to the staff that the patients did not belong in the hospital. This did not happen. In fact, something very frightening happened!

The amount of time the patients stayed hospitalized varied from 7 to 52 days. No psychiatrists or hospital staff members suspected that the patients had faked their symptoms to gain admission to the hospital. Nursing reports described them as friendly and cooperative, but never suggested that the fake patients did not belong in a mental health facility.

The pseudo patients often took notes, which they feared would make the staff suspicious. On the contrary, the constant note taking was described as a symptom of the patients' mental illness.

An interesting part of this study was that many of the real patients in the mental hospital suspected the fake patients of being journalists and fakes.

The inability of the mental health staff to recognize normal behavior while many of the actual patients were able to, was especially disturbing and shows that confirmation bias is deeply embedded in mental health institutions. The behavior of the pseudo patients was constantly filtered through the lens of expectations, influencing how the actions and words of the fake patients were interpreted.

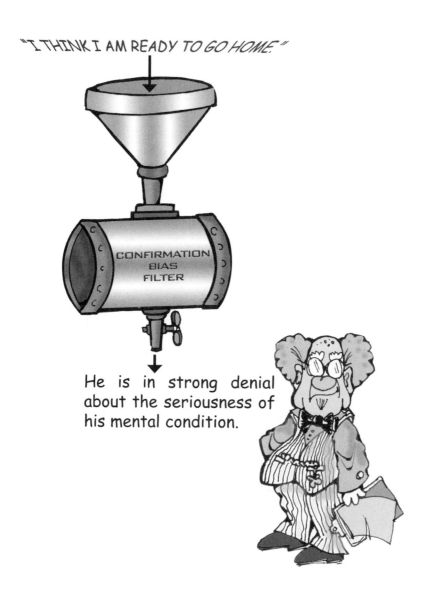

An example of this filtering process occurred when one of the pseu-do patients reported that he had a close relationship with his moth-er as a child, but that it noticeably cooled when he became an adult. He also mentioned that while he was a child he was distant from his father, but became closer to him as he entered adulthood. The fake patient reported that he had a warm and close relationship with his wife with occasional arguments, but little friction. He also reported a rare need to spank his children.

The mental health professional used this information to write the following report:

> *"The white 39-year-old male...manifests a long history of considerable ambivalence in close relationships, which begins in early childhood. A warm relationship with his mother cools during his adolescence. A distant relationship with his father is described as becoming very intense. Affective stability is absent. His attempts to control emotionality with his wife and children are punctuated by angry outbursts and, in the case of the children, spankings. And while he says that he has several good friends, one senses considerable ambivalence embedded in those relationships also..."*

The facts of the pseudo patient's life were clearly altered to match those of someone who suffered from mental illness. This study shows how difficult it is to overcome a label of mental illness once it has been diagnosed. All behavior --- even normal behavior --- is put through the confirmation bias filter and rearranged until it matches what is expected of a mentally ill individual. This is especially frightening because a diagnosis of mental illness, whether it is correct or not, can forever stigmatize a person's emotional, social, and legal life.

There is an interesting postscript to this story. After the hospital staff was told about the fake patients, many of them doubted that their hospital could be fooled so easily. Rosenhan then told the hospital staff that at some point in the future he would repeat his experiment and asked the staff members to take notes on future patients whom they thought were not genuine. The results were very interesting:

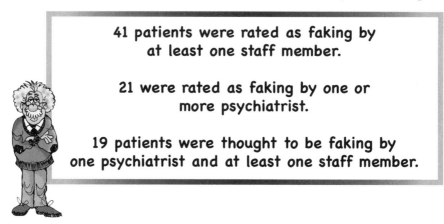

**41 patients were rated as faking by
at least one staff member.**

**21 were rated as faking by one or
more psychiatrist.**

**19 patients were thought to be faking by
one psychiatrist and at least one staff member.**

How many fake patients did Rosenhan send in the second round of his experiment? NONE! Every patient that the staff judged as fake was in fact a genuine patient.

The results of these two studies attest to the power that bias can have over our judgment. When we expect to see a mentally disturbed patient, then every behavior from him is influenced by that expectation, which affects how we interact with and view him. Confirmation bias is not simply a laughable phenomenon that makes people believe weird things like cats predicting death or paintings containing hidden musical messages, it is also a dangerous by-product of our brain's architecture that can lead to prejudice and misunderstandings. By focusing on what we want or expect to see and ignoring the rest, we risk living in a reality masked by our own expectations.

Facilitated Communication

Autism is a profound developmental disability that involves an impaired ability to relate to other people. In addition, autistic children often suffer from language difficulties and often rigidly adhere to routines.

During the 1980's, an interesting discovery came out of Australia. When adults held the fingers of mentally disabled autistic children over keyboards, the children typed out messages that were far beyond their supposed abilities. Children as young as 6 or 7, who were thought to have little or no writing or reading skills, typed messages that were more typical of children much older. Almost everyone who saw this occur was stunned because it appeared that not only were these children not mentally disabled, but they seemed to be extraordinarily gifted at communication.

This was indeed a promising development for the autistic community. It now appeared that these children simply had difficulty expressing their thoughts. Once they were assisted by an adult at a keyboard, they were able to communicate.

In the United States, a small number of educators heard about this new technique and tried it on the autistic children they were working with. The educators held the fingers of the children over keyboards and, just as in Australia, the American children appeared to type messages that shocked their teachers. Children who were previously thought to possess no reading or writing skills, were not only typing complete sentences, but were also expressing thoughts that required advanced reasoning ability.

As word of this miracle spread across the country, thousands of autistic children received keyboards and adult facilitators. In addition, education plans were dramatically changed because the typing that appeared to originate from the children made it clear that they were functioning at a higher intellectual level.

This new technique was called facilitated communication and it spread rapidly across the United States. Unfortunately, this occurred before any scientific tests were done to see if it really was the children who were communicating through the keyboard. Then something happened that made it imperative that testing be done - **messages appeared on computer screens that accused parents and others of abuse.**

Fathers were jailed and children were taken away from their parents based on messages that were typed on keyboards. As the accusations of abuse entered the judicial system, the courts demanded that the question of who was communicating be determined scientifically.

> What is so sad about this story is that it is very easy to scientifically test facilitated communication. All that needs to be done is to show the child and the adult different pictures and then see what is typed on the keyboard.

When the scientific tests were completed, the results were shocking. No matter how many times pictures were shown to the adult and the child, the keyboard entry was invariably the name of the picture the adult saw. These results made it clear that the adults were the authors of the communication, not the children.

What was even more shocking than the results of the scientific experiment was the reaction of many of the people who were using facilitated communication. They ignored the scientific evidence and continued to believe that it was the children who were actually communicating through the assisted typing.

The power of confirmation bias is documented by the reaction of those who believed strongly in facilitated communication. They quickly and easily ignored what math and science told them. They maintained their strong belief even though the autistic children often were looking out a window during the typing, while the adults had their eyes glued to the keyboard. It was almost as if the fact that the children were communicating was a given and all contradictory evidence had to be explained to fit that belief.

There was even a case where a Spanish speaking facilitator was working with a young autistic child who did not know Spanish. Much to the surprise of the boy's teachers, the communication that appeared on the computer screen was in Spanish. Instead of acknowledging the obvious --- that the communication on the computer had to be coming from the Spanish speaking adult facilitator --- it was decided that the autistic child had the uncanny ability to learn a language by just sitting next to a Spanish speaking individual. **Now that is confirmation bias!!!!**

The educators really should have known to test facilitated communication scientifically before they started using it. At least the courts required scientific proof and freed the falsely accused people.

"The human understanding, once it has adopted an opinion, collects any instances that confirm it, and though the contrary instances may be more numerous and more weighty, it either does not notice them or else rejects them, in order that this opinion will remain unshaken."
--- Philosopher Francis Bacon (1620)

CHAPTER ELEVEN
Straw Man Arguments

Healthcare Debate Tactics

In the years 2009 and 2010, President Obama and the United States Congress undertook an effort to reform the country's health care system. The proponents of healthcare reform had many reasons for their proposed changes, and opponents of these changes had many legitimate concerns that had to be debated and discussed to ensure that the bill that came to the President for his signature was the best bill possible.

Concerns of Opponents
- National debt might skyrocket because of the cost of reforms
- Too much government involvement in health care
- Concerns about requiring people to buy insurance

Changes Wanted by Proponents of Healthcare Reform
- Need to insure every American
- Keep health care affordable
- Stop the practice of denying insurance for pre-existing conditions

Moderator:

Looks like we have a great debate about to start. There may be some honest differences of opinion, but it appears both sides have some good ideas. After some intellectual sparring and a rational discussion of the facts, I'm confident that the best bill possible will emerge. Okay, it looks like the opponents of the healthcare plan are going to go first.

"The America I know and love is not one in which my parents or my baby with Down Syndrome will have to stand in front of Obama's 'death panel' so his bureaucrats can decide, based on a subjective judgment of their 'level of productivity in society,' whether they are worthy of health care. Such a system is downright evil."

A straw man argument or debate strategy occurs when one side misrepresents an opponent's position and then attacks and destroys that false position. The actual position of the opponent is not debated, but the straw man position is attacked and easily defeated. Victory is declared without the rigor of an intellectual challenge.

The problem with the "death panels" straw man argument in the healthcare debate is that there was never a discussion of adding death panels to the healthcare bill. As a matter of fact, there was no discussion of anything resembling death panels. What was discussed was the very reasonable subject of end of life counseling for terminally ill patients. Anyone who has cared for a terminally ill relative knows that end of life counseling is a reasonable part of any healthcare reform discussion.

A fair analogy to the death panel straw man argument used by opponents of healthcare reform would be the following:

I think it is important that people consult with their veterinarian about the available options when their pet is nearing the end of its life.

I will not have a government panel decide whether my dog is allowed to live!!

A straw man argument is usually used by a desperate opponent who is losing an argument or cannot hope to honestly debate an opponent's actual position. A straw man argument is a dishonorable approach when discussing an issue or point of view, but it is used not only in politics, but also in all walks of life.

We have just seen how some Republicans use straw man arguments to redefine an opponent's position to make it weak and easier to debate. Of course Democrats also use the same debating technique.

During a speech, President Obama proclaimed that there was a showdown between himself and those who wished to do nothing about the recession. "Nothing is not an option. You didn't send me to Washington to do nothing."

> Contrary to what the President said, Republicans did have many plans to deal with the economic meltdown. Among them were suspending payroll taxes, slashing income tax rates and lowering the tax on capital gains.

It certainly would have been reasonable to debate the Republicans in a direct, honest way about their proposed solutions to the country's economic woes, but using a straw man argument promoted conflict and a poor understanding of the issue. When you use straw man arguments, you engage in a debate with no intellectual depth.

The 2010 Coal Mine Explosion
and the CEO's Straw Man Defense

In April of 2010, a deadly explosion occurred in the Upper Big Branch coal mine killing over 25 workers. The company running the mining operation had been cited for over 600 violations in the previous 18 months, including several for improperly ventilating methane, the most likely cause of the explosion.

The disaster put the safety record of Massey Energy, the company that runs the mine, under a microscope. Experts said that while it is common for a mine as large as Upper Big Branch to have many violations, most don't have the number of serious infractions that the Upper Big Branch collected. They had been cited over 50 times for failure to comply with safety standards such as:

• Failure to follow an approved ventilation plan

• Failure to control combustible material

• Not designating an escape route

In addition, the CEO of Massey suggested in 2006 that two supervisors be fired because they expressed safety concerns about conveyer belts in the mine. Shortly thereafter, a conveyer belt caught fire and two miners were killed.[25]

Now the CEO of Massey Energy has a devastating explosion at one of his facilities. How can he possibly defend his mining operation in light of the large number of deaths and the company's long list of safety violations?

Massey CEO:

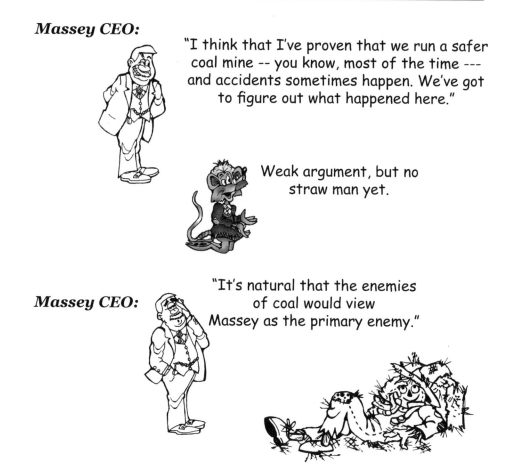

"I think that I've proven that we run a safer coal mine -- you know, most of the time --- and accidents sometimes happen. We've got to figure out what happened here."

Weak argument, but no straw man yet.

Massey CEO:

"It's natural that the enemies of coal would view Massey as the primary enemy."

There it is --- a classic straw man. The CEO's attitude is that people who take Massey Energy to task for hundreds of safety violations, many of them serious, will not be engaged on the issues at hand because the Massey CEO really has no defense. He then resorts to attacking a straw man.

We have a mining operation that has been cited numerous times for failure to ventilate methane properly, and then a deadly explosion occurs killing 25 miners–possibly because of improperly ventilated methane–and the CEO rails against the critics as being enemies of coal.

Conclusion:

Arguing with a straw man is easier than really debating. Look at the following examples and see how easy it is:

Iraq War

Liberal: Your answer to everything is war!

Conservative: Why do you hate America?

There were many arguments for invading Iraq and many for not invading Iraq. Neither of these statements contributes to a real debate on the issue.

Cuts in the Military budget

Conservative: Why do you want to leave the country defenseless? Also, why do you hate America?

Liberal: Why does the military get every weapon system it wants?

Education

Conservative: I don't like the changes our education system has made recently.

Liberal: Some people are afraid of all change and always want the status quo.

Marriage

Woman: Honey could you take out the trash?

Man: Why does it bother you when I relax?

Children and Dad

Dad, can we get a dog? ....

　　　　　　　　　　... No

It would protect us. ...

　　　　　　　　　... No

Why don't you care about our safety? ...

　　　　　　　... Sorry, I don't engage in straw man arguments

Remember, those who are incapable of defending a position or belief argue for something that is easy to defend and pretend it is the actual position or belief that is being debated.

CHAPTER TWELVE
Manipulating Mean, Median and Mode

Then there is the man who drowned crossing a stream with an average depth of six inches.

--- W.I.E. Gates

The difference between mean, median and mode is almost always the first topic of discussion in any course in statistics. This is because many of the more advanced statistical functions such as the ANOVA (analysis of variance), factor analysis, and standard deviation can only be understood by knowing what averages are and how they are created. Rarely taught, however, is the fact that means, medians and modes, can be used in different scenarios to demonstrate certain trends. As a result, there are situations when it is better to use each form of average in order to fairly represent data and avoid deception.

In order to understand how mean, median and mode are misused, it is very important to know the difference between the three types of average.

Mean: The mean is the most common method of averaging. To find a mean, you add up all the data points and divide by how many points there are. For example, if 5 farmers have 2, 3, 4, 7, and 8 chickens, to find the mean number of chickens you would add 2+3+4+7+8 and then divide the total by 5, resulting in a mean of 4.8 chickens. When someone uses the word "average" but does not specify which type, it is generally understood that the "mean" average was used, but not always.

Median: The median average of a set of numbers refers to the number in the middle. This can be found by arranging the numbers in ascending order and then selecting the number in the middle. Using the above example, the median would be 4 chickens, because 2 farmers have fewer than 4 chickens and 2 farmers have more than 4 chickens.

$$2, \ 3, \ \mathbf{4}, \ 7, \ 8$$

When you need to find the median of an even set of numbers where there is no clear middle, you select the middle two numbers and find the mean average of these. For example, if one more farmer came along and had 12 chickens, you would add 4+7 (the two numbers in the middle) and divide by 2, giving you a median of 5.5.

> 2, 3, **4, 7**, 8, 12

Mode: The mode is the easiest form of average to understand. Basically, the mode is whichever number occurs most in a sequence. It is also possible that a set of numbers has no mode, such as in the above example because each number is different. The mode is rarely used as a form of average simply because it does not summarize the data as well as means and medians do. For example, the mode in the following data set would be 2:

2, 2, 100, 500, 465, 783, 234, 645, 243

While this is the most commonly occurring number, it does not tell us anything about the remaining numbers, as the other two forms of averages would. Where the mode shines through, however, is when we are trying to find the average of something non-numerical. For example, if you wanted to paint your bike a color that would fit in with the rest of your friend's bikes, you could find the average color by using the mode:

> **Blue, red, red, white, brown, blue, black, blue, orange.**

Because you cannot add or divide colors as you can numbers, nor can you arrange them in increasing order, the only way to find an average of data such as these would be to find the mode. In this case, the average bike color is blue.

Each method of finding averages has its own advantages and disadvantages when working with numbers. When the data is fairly symmetrical and does not have extreme outliers, a mean average should be used. On the other hand, when there are a small number of outliers that heavily skew the mean, a median should be used to accurately represent the data. An honest reporter will use each type of average to demonstrate the facts in the most accurate and truthful way; however, dishonesty can also cause people to intentionally use a certain average to distort the truth.

The most common way to demonstrate this is to look at income. Because incomes can range from 0 to billions of dollars, there is great potential for the mean of incomes to be skewed in one way or another. The top 1% of earners in the United States have a much higher proportional income than the remaining population, and therefore have a tendency to increase average incomes when the mean method is used.

For example, if you live in a suburb that has a total of 100 people, all having an income of around $50,000 a year, the mean would be around this number. However, if Lebron James, with an NBA salary of 15.7 million dollars a year, decides to buy a house there, the mean annual salary would go from $50,000 to around $205,000.

It is because of outliers such as these that using medians can more honestly portray average incomes. In this example, the median average would not be affected even if Lebron James (15.7 million), Kobe Bryant (23 million), and Shaquille O'Neal (20 million) all decided to move into the area. The middle, or median income would still be around $50,000. (Even though the mean income would be around $620,000.) In any example, finding the mean average of a set of numbers has the potential to be very misleading if there are any outliers to the data.

Unfortunately, the abuse of averages remains rampant in this country. Whether this is always intentional or not can be hard to prove, but when someone chooses to report a mean instead of a median and makes the numbers look more favorable to their position, there is a high probability that they are aware that deception is involved.

For example:

In promoting the tax cuts of 2003, the White House claimed that 92 million taxpayers would receive an average tax cut of $1083. However, according to the Urban Institute-Brookings Institution Tax

Policy Center, 80% of taxpayers would receive a tax cut less than this average number. Because 80% of taxpayers would receive a tax cut below the $1083 average, the White House's estimate was clearly an average based on the mean.

The $1083 figure is misleading because while the word "average" suggests it represents the majority of Americans, the average in this case represents a small minority of taxpayers. In fact, according to the Tax Policy Center, 49% of taxpayers would receive less than $100 in tax cuts, less than a tenth of the "average" tax cut. Also, the mean tax cut for the bottom 80% of taxpayers came to a mere $226. The top 1% of earners in the country balanced these low averages and made the total mean disproportionately higher. This top 1% alone received an average of $24,100 in tax cuts, and people who earned above a million dollars a year received tax cuts of around $90,200.

The large tax cuts for the top earners in the country greatly inflated the mean average to the misleading $1083 figure that the White House asserted. The word "average" implies that it represents the majority of the population; however, when reporting the mean average of data such as income, which has a wide range, the results will almost always be skewed one way or another as those with a disproportionate income disproportionately affect the numbers. In this case, reporting the median would have been more honest, as it would demonstrate the middle tax cut for all Americans, and therefore would represent that of the majority of all taxpayers. [26]

Using mean averages when talking about controversial issues such as taxes is a popular type of statistical manipulation because of its "technically accurate but misleading" nature. George W. Bush wasn't lying when he stated in his State of the Union address that if his tax cuts were allowed to expire, then "116 million Americans would see their taxes rise by an average of $1,800." Although not a lie, this figure remains misleading because it is highly skewed by the top 1% of taxpayers in the country.

To put this into perspective, the Washington Post checked the facts on the State of the Union address and found that "the median American household will pay roughly $828 more in taxes in 2011 if the Bush tax cuts expire, according to the Tax Policy Center, a non-ideological think tank venture. The richest 1 percent of American households, in contrast, would have to pay an extra $64,154 a year when the tax cuts expire."

While people would also certainly object to paying $828 more in taxes for the year, this figure is more representative of the majority of Americans, and also less than half of President Bush's $1,800 assertion. The NPR News Blog phrased the tax cuts in another way:

> *The Citizens for Tax Justice estimate that the middle 20 percent of Americans will receive 11 percent of the Bush tax cuts between 2001 and 2010, while the top 1 percent will receive 36 percent. That means the middle 20 percent would lose about $540 a year in tax breaks if the Bush tax cuts are not renewed. The top 1 percent would lose an average of $34,000 a year.*

The analyses by the Tax Policy Center and Citizens for Tax Justice show how topics as complicated as taxes cannot be expected to make much sense by using a simple arithmetic average. Whenever anyone uses the word average when talking about income or anything income related (such as taxes), be aware figures can be skewed because the top 1% of the population earns an amount disproportionate to the rest of the country.

For more information see the following websites:

http://www.npr.org/blogs/news/2008/01/
in_the_speech_we_have.html

http://www.washingtonpost.com/wp-dyn/content/
article/2008/01/28/AR2008012803175_pf.html

http://mediamatters.org/research/200801290012

The next chapter shows how the compelling nature of testimonials and anecdotal evidence has lead to thousands of people basing medical decisions, including whether or not to vaccinate their children, on advice that is promulgated by celebrities who have little or no understanding of how scientific inquiry works.

CHAPTER THIRTEEN
Anecdotal Evidence

Anecdotal evidence is information that is derived from a limited amount of data, which usually comes from personal accounts (testimonials) and observations. Anecdotal evidence is usually not supported by scientific studies, but because it can be very convincing, it is used by a large number of people to form beliefs and make decisions.

Look at the following examples that illustrate the difference between anecdotal and scientific evidence:

This is scientific evidence:
Thousands of people who took Vioxx were compared to thousands who took Aleve. Those in the Vioxx group had four times the number of heart attacks as the people in the Aleve group. From this information it was determined that there is a high probability that Vioxx causes heart attacks.

This is anecdotal evidence:
My Uncle Steve took Vioxx, had a heart attack and died. Vioxx causes heart attacks!

These examples not only highlight the clear differences between scientific and anecdotal evidence, but also hint at the value of anecdotal evidence when it is used properly.

While anecdotal evidence does not equate to proof, it does not mean that this type of evidence is worthless. Anecdotal evidence about the dangers of tobacco, asbestos, and Vioxx led to the scientific studies that determined that these substances are harmful.

In addition to indicating areas that may need scientific investigation, there are other ways anecdotal evidence can be used that are reasonable and helpful to the decision-making process.

If you want to know the best electrician or plumber available, what doctor to see, or what restaurant to patronize, it is very unlikely that you will be able to base your choice on scientific testing. What you would probably do is ask friends about their experiences and then make your decision based on the limited anecdotal evidence that you gather.

The use of anecdotal evidence becomes problematic, though, when it is used to make decisions that affect our health and financial situation.

Illness and Anecdotal Evidence

One of the reasons alternative medicine has grown in popularity is that a certain percentage of patients will get better without any treatment, and therefore, any "medicine" that is used appears to be involved in the cure. Even the most ridiculous claims of some proponents of alternative medicine have their own "evidence" in the form of anecdotes or testimonials.

The problem with anecdotes, as they apply to health issues, is that they are almost always never first-hand stories; instead, they involve something that happened to a friend-of-a-friend. Second, when stories are told over and over again, the details become exaggerated and memories get distorted.

Even first-hand testimonials cannot always be trusted because our own biases or previous memories frequently influence our interpretation of events. These misinterpretations are not always deliberate, but may be a result of confirmation bias, the placebo effect, or a combination of other causes. In addition, some stories are intentionally fabricated, some are a product of delusions, and some events may be deemed supernatural simply because their scientific basis is not understood.

Before we understood the scientific basis of how germs cause disease, the ideas for what caused an illness ranged from bad vapors to witchcraft.

Regardless of their accuracy, compelling stories spread through populations like wildfire, and as a result testimonials simply cannot be treated as facts and have little (if any) scientific value.

Unfortunately, because of the compelling nature of testimonials, thousands of people now base medical decisions, including whether or nor to vaccinate their children, on advice that is promulgated by celebrities who have little or no understanding of how scientific inquiry works.

Despite their fallibility, anecdotes remain a popular method of persuasion. Television commercials often rely on this method by showing a celebrity or sports star who has (allegedly) tried a product and encourages others to do so.

When hearing a testimonial, the motivation of the speaker is important to discern before giving the story any credibility. Advertisements offer a clear motivation: money. Does Shaquille O'Neal really use Icy Hot when he has sports pain? It's possible, but the fact that he receives large sums of money to use it in front of a camera tells us more than his testimonial itself does.

One of the reasons anecdotal evidence is hard to ignore, even when there is strong contrary scientific evidence, is that our brains are programmed to look for patterns (see chapter 10: Confirmation Bias). As mentioned in the earlier chapter, it is much less dangerous to our survival to see a pattern where none exists, than to miss a pattern where one does in fact exist.

Magnetic Therapy

There is a debate among believers of magnetic theory as to how it works, but most advocates agree that placing magnets on the body somehow increases circulation. Sites explaining the therapy use scientific words, such as "the magnetic field energizes and oxygenates the white corpuscles in the blood stream, and these white corpuscles are natures [sic] healing agents." This website sells a complete list of magnetic products, including magnetic pillows, car seats, patches, travel pads, adhesives, and discs, showing that there is a large profit potential for a therapy that is completely devoid of any scientific evidence.

I had surgery to remove painful nerves in my left hand, and it left me with bad circulation and numbness. I tried a magnetic bracelet and in about an hour I could feel the ends of my fingers again. I won't take it off!!

Testimonial:

When I heard about magnets, I had been living with pain for about a year. My forearm was crushed in a hydraulic lift gate, severing a tendon and breaking the bone. It didn't heal right and I had constant twinges of nerve pain throughout the arm, especially after using the computer. Within about 20 minutes of wearing a magnetic bracelet, the pains stopped, and now I wear one most of the time. If I go longer than 24 hours without my bracelet I can tell a difference. I have friends who wear the bracelets and no longer have trouble from carpal tunnel syndrome. My mom swears the magnetic insoles keep her feet from feeling achy and burning at the end of the day.[27]

Note that this testimonial included not only the primary believer, but also "friends" and "my mom", in an attempt to reduce skepticism by adding to the list of believers.

Dowsing

Dowsing often involves using a simple instrument (such as a stick or metal rod) to locate buried water, metals, or oil. Most commonly, dowsers hold Y-shaped branches and walk slowly over suspected areas, looking for any dips or twitches their "instrument" may make. Another method uses two L-shaped metal rods. People carry one in each hand, and when they cross over each other and form an X, it supposedly means the targeted object is below the dowser.

Dowsing advocates disagree on many specifics, such as certain types of wood or metals being more effective than others. While the disparity among its followers should be enough in itself to suggest how fallible the practice is, dowsers have also failed to achieve a higher rate of success than chance in controlled laboratory experiments.

Even though dowsing does not hold up to scientific scrutiny, this doesn't deter people from believing wholeheartedly in the phenomenon. The following information is from the testimonial section on a website that sells "Teletherapy," which claims to use a dowsing pendulum to transmit "healing energies to a person located at a distance from the healer or the healing instruments."

According to the website, "The Phenomenon has been proved by Dr. Sarkar, of Calcutta, India, about twenty years ago. He was a practitioner of Color and Gem therapy and an expert dowser. He used a photograph of a patient located in New York to radiate red color light to the patient while Dr. Sarkar was in Calcutta. Two independent doctors were observing the patient through a prism in New York. During the fifteen-day experiment, the observing doctors recorded a significant increase in the red color energy of the patient day by day."

The claims produced no sources or literature, but in its place they provided plenty of testimonials written by alleged patients. An example of a testimonial is shown on the next page.

Testimonial:

My daughter was pregnant and the doctor told her to abort the child because some medical reports suggested that the baby might have some genetic fault. I consulted my Modern Vastu (Energy) Science guruji Sri. K. H. Betai and he dowsed and informed us not to worry at all as the baby was normal as per dowsing. My daughter delivered a normal baby girl in December 1998 in USA.[28]

Alien Encounters

There are an abundance of tales from people who claim to have had an encounter with an alien. Although the stories have an enormous variety, ranging from fleeting glimpses to abductions, the consistent theme of all reported encounters is that they have never produced any physical evidence.

One of the two testimonials shown below is an actual report of an alien encounter, the other is fabricated. See if you can tell the difference.

> *He and a teen friend were being driven by his father past the Cleveland swamp when they saw a luminous UFO land in the swamp. They went on to their destination (the next town) and when returning by the same road an hour or so later, saw a "7-foot-tall alien by the side of the road right near where we saw the UFO go into the swamp." We wanted to stop, he said, but my dad was too scared and stepped on it instead.*[29]

> *I woke up in the middle of the night with the feeling that something else was in the room. There was a shadowy figure in the corner of my room that was about four feet tall with large eyes and gray skin. There was a flash of light and I woke up a few hours later with a fever and no memory of what just happened.*

The first story is the real testimonial about an "alien encounter" while the second story was made up. Because inventing testimonials is as easy as telling a story, something that a large percentage of the world population is certainly capable of, there is no way to verify the accuracy of a testimonial by reading it alone. Many of these stories would be easy to prove, but as mentioned earlier, of the thousands of alleged alien abductions, not one has provided us with a piece of physical evidence.

Unfortunately, even though it would be very easy to prove the existence of aliens, it is almost impossible to prove these stories false because a number of fantasy-based explanations can be invented to answer any question based on logic. In *The Demon-Haunted World*, Carl Sagan illustrates this in his example of how hard it would be to prove that an invisible dragon doesn't live in your garage:

"You propose spreading flour on the floor of the garage to capture the dragon's footprints.

'Good idea,' I say, 'but this dragon floats in the air.'

Then you'll use an infrared sensor to detect the invisible fire.

'Good idea, but the invisible fire is also heatless.'

You'll spray-paint the dragon and make her visible.

'Good idea, except she's an incorporeal dragon and the paint won't stick'"

Because anecdotes are not based on physical evidence, contradictions among these stories can be explained with a number of nonsensical clarifications. As such, stories by themselves cannot be used as proof for anything. However, they are far from useless and, as mentioned earlier, can guide scientific research toward verifying their claims.

It is important to remember that when you are trying to decide how credible an anecdote is, you must remember to take into account whether the person telling the anecdote has anything to gain (such as money or attention).

Conclusion

Paranormal claims such as psychic abilities or dowsing rods rely on anecdotes because they cannot be verified through controlled tests. Unfortunately, despite the lack of scientific evidence, people often misinterpret enthusiastic testimonials as facts and base their decisions on them.

A useful tool when sifting through an abundance of testimonials is Occam's Razor, which teaches that the simplest explanation is usually the most likely one. In other words, when you hear hooves, think horse, not zebra. Occam's Razor is also very helpful when the latest Bigfoot sighting is reported on the local news. Occam's Razor tells us that the most likely explanation for Bigfoot is not that a new species of very tall human-like creatures has been discovered. Occam's Razor would guide us towards too much alcohol, drugs, or a prank as the explanation of why someone reported seeing Bigfoot.

Simply put, testimonials and anecdotes are not reliable when compared to scientific evidence, which can be replicated, challenged, and disproved.

CHAPTER FOURTEEN

If You Don't Have Statistics to Support Your Cause, Just Invent Some

"Statistics can be made to prove anything - even the truth."
--- Author unknown

The Power of Subliminal Advertising

In the 1950's, a market research consultant named James Vicary announced the results of a six-week experiment he conducted that documented the power of subliminal advertising. During the experiment, Vicary flashed messages onto a movie screen during showings of the movie *Picnic* that said **"Drink Coca-Cola"** and **"Hungry? Eat Popcorn"**. Each message, which lasted 3/1000 of a second, was far too short to be consciously perceived by the brain.

The results of the experiment were stunning. When **"Drink Coca-Cola"** was flashed, sales of Coca-Cola increased 18 percent and when **"Hungry? Eat Popcorn"** was flashed, sales of popcorn increased by 58 percent!

Because of Vicary's experiment and the subsequent media attention devoted to it, the concept of subliminal messaging is embedded in our culture. Psychology classes and advertising classes devote attention to it, and countless books and articles have been devoted to the subject.

> One of the wonderful things about science is that when a new discovery is made, others in the field try to duplicate the results. It is a very effective check against experimental error and fraud.

Were others able to recreate Vicary's experiment that showed the power of subliminal advertising to influence behavior? In a word, no. The reason that the experiment could not be duplicated was not because the experiment had design flaws or that Vicary manipulated the data he generated. The reason that others could not show an effect from subliminal messaging was that James Vicary never did his ground-breaking experiment --- he made the whole thing up!

Unfortunately, more than 50 years later, it is still a generally accepted fact that subliminal messaging can influence the behavior of individuals.

The Unusually High Mortality Rate from Anorexia Nervosa

Anorexia nervosa is an eating disorder that usually occurs in young women and is characterized by very low body weight, a distorted body image, and a fear of gaining weight. Concerned activists, in order to make the public more aware of the dangers of anorexia nervosa, warned that the approximately 150,000 American women who have the eating disorder are at risk of dying if they are not treated.

At some point, the 150,000 women suffering from anorexia turned into 150,000 American women **dying** each year from the eating disorder. This dramatic new statistic was repeated in books, magazines, newspapers and on news shows. Once the incorrect "150,000 deaths from anorexia per year" appeared in print, it was very easy to source the information and the spread of the inaccurate statistic was almost impossible to stop.

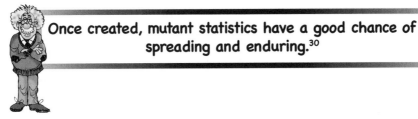

Once created, mutant statistics have a good chance of spreading and enduring.[30]

The change from 150,000 cases of anorexia to 150,000 deaths was probably not done deliberately (although 150,000 deaths per year from anorexia garnered a lot more attention than the more accurate 50-100 deaths per year would have). The most disturbing aspect of this story is that the hundreds of people who repeated the inflated mortality rate should have known better --- fewer than 10,000 women in the 15-25 age group die each year from all causes. 150,000 deaths a year from an eating disorder just does not make sense.

Did you know that each year 150,000 women suffer from anorexia.

I just heard that 150,000 women die each year from anorexia!

3 Million Homeless in the United States

There have been many advocates for the homeless over the years. One of the most enterprising and dedicated of these was a man named Mitch Snyder who traveled the country in the 1980's trying to bring the plight of the homeless to the public's attention. He accomplished this by involving the media in the issue and in the process did countless interviews and media events.

When Mr. Snyder was asked to estimate the number of homeless in the United states, he would respond that it hovered around 3 million. The media then used the 3 million figure and it was repeated so often that it became widely accepted as fact. The media continued to use the "3 million homeless in America" statistic until Ted Koppel confronted Snyder about the source of his estimate. Snyder eventually admitted that he essentially made up the number in response to media inquiries.

A more accurate estimate of the number of homeless in the United States in the 1980's is 300,000 - 400,000. This does not make the problem any less serious for those affected, but 300,000 homeless is a much different issue than 3 million homeless.

 What number of homeless do I need to give to make you pay attention to the problem?

There is a constant stream of data coming from interest groups with causes to promote. Mitch Snyder's fabricated 3 million homeless statistic shows how important it is for the media and the public to demand documentation and accuracy when advocacy groups give numbers.

Other Erroneous Statistics
That Have Made it Into the Popular Lexicon

DAILY NEWS

100,000 dogs are killed each year falling out of pickup trucks.

275 dogs per day? Not likely!

DAILY NEWS

There is a 40% increase in domestic violence during the Super Bowl.

Actually, there is no change in domestic violence during the Super Bowl.

DAILY NEWS

Abuse of pregnant women is the number one cause of birth defects in the United States.

This came from a " March of Dimes" study that turned out not to exist. Of course spousal abuse is common in the United States, and it certainly could cause injuries to an unborn child. However, it is just not likely that spousal abuse could be responsible for more birth defects than alcohol, drugs, AIDS, and other genetic disorders such as Down Syndrome. (For more information about fabricated statistics, read *Damned Lies and Statistics* by Joel Best.)

CHAPTER FIFTEEN
Perplexing Percentages

The Doctors Who were Fooled by Percents

Because it is very easy to be fooled by statistics, millions of people are misled each day by information they read. This can have serious consequences, but it is especially dangerous when doctors are misled by statistics. A group of scientists decided that it would be interesting to test doctors to see if they could tell how effective a drug was by looking at statistics on that drug.

The scientists gave 50 doctors the results of a study about a heart drug where .3% of the patients getting the drug had fatal heart attacks, while .4% of the patients who didn't get the drug had fatal heart attacks. Because there was only a .1% difference between the number of deaths in the two groups, the doctors were not very excited about the drug.

The scientists then gave 50 different doctors the same information about the heart drug, but they presented it in a slightly different way. They told the doctors that the heart drug caused a 25% decrease in deaths. The doctors who were given this information were much more likely to say they would prescribe the drug than the first group.

Even though it doesn't appear to be the case, the information that the researchers gave each group of doctors was identical. A drop from .4% to .3% was the same as a 25% drop. This experiment showed that the way numbers are presented can influence what people believe --- even highly educated doctors!

Remember that the way to determine a percent of change is to take the amount of change and divide it by the original.

I bet pharmaceutical companies know how to present data to doctors.

The Growing Trend of Children Living at Home

An article in USA Today (2006) suggested that there was a growing trend of young adults living at home.

Quote from the newspaper article:
"Since 1970, the percentage of people ages 18 to 34 who live at home with their families increased 48%, from 12.4 million to 18.6 million."

The article discussed several possible reasons for the increase, such as high housing costs.

A 48 percent jump in young adults living at home is significant and possibly a sign that economic conditions are having much more of an impact on young adults than previously thought. But before we start building additions to our homes for returning children, let's give some meaning to the provocative statistics.

The population of the United States was 204 million in 1970, so at that time, 12.4 million divided by 204 million or 6.1% of the population were young adults (18 to 34) who lived at home. In 2006, the population of the United States was 297 million, so 18.6 million divided by 297 million or 6.3% of the population were young adults (18 to 34) who lived at home.

In 1970, 6.1% of the population were young adults living at home. Thirty-six years later, 6.3% of the population were young adults living at home, an increase of .2 percentage points as opposed to the reported 48% increase --- not much of a trend. (The percent of increase is .2 ÷ 6.1 = 3%)

Struggling Children of Immigrant Families

A recent New York Times editorial (11/16/2009) discussed the academic struggles of immigrant children. Embedded in the thoughtful and heartfelt writing was data that was clearly meant to highlight how immigrant children were failing to properly master standard English.

"Whereas native-born children's language skills follow a bell curve, immigrant children were crowded in the lower ranks: More than three-quarters of the sample scored below the 85th percentile in English proficiency."

The editorial went on to relate the very real stumbling blocks facing these children when they fail to master English, which is --"the passport to college and to a brighter future."

The data presented in the article is either incorrect, or if correct, can lead us to only one conclusion --- immigrant children are more proficient at English than native born children. If you take a test and score at the 85th percentile, then you scored better than 85 percent of the others who took the test.

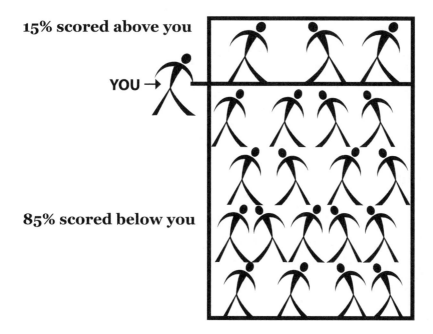

15% scored above you

YOU →

85% scored below you

To understand this better, let's look at a standard bell curve for a test:

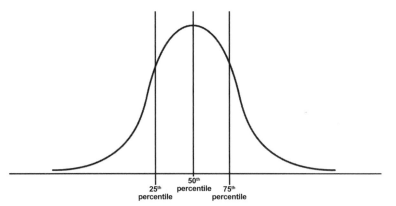

If you score at the 50th percentile, you did better than half of the other people who took the test. If you score at the 75th percentile, you did better than 3/4 of the other people who took the test.

Now, if we want to see how certain groups do on the test, we can compare the group's results to the population as a whole. If a certain population, say left handed people, took the test and 50% scored below the 50th percentile, then we can say they are pretty typical of the group as a whole. But if 80% scored below the 50th percentile, then we know this group is struggling compared to the population as a whole.

On the other hand, if 40% of the left handed group score below the 50th percentile, then we know left handed people are doing better than the population as a whole because 50% didn't fall below the 50th percentile --- only 40% of the group did.

Let's take the group of immigrant children. If 85% fell below the 85th percentile, then we could say their testing is pretty typical of the general population. But only 75% fell below the 85th percentile. This means they did better on the test than the population as a whole.

The data in the editorial does not match the meaning of the article. I hope the author knows what percentile means.

The Mysterious 500% Drop in Demand

Hastings Minnesota
DAILY NEWS

Since inmates at the Dakota County Jail have had to pay for over-the-counter medications, there has been a 500 percent decrease in demand.

Let's say the demand for over-the-counter medications was 1000 bottles and it dropped to one bottle. The percent of decrease would be 999/1000 or 99.9%. You can see from this example that the largest possible percent of decrease is 100%.

There are some very odd situations where one could claim a percent of decrease that exceeds 100%. If a painting originally cost $100 and now the sellers pay you $400 to take it away, the change is $500, divided by the original price of $100, or 500%.

I Never Should Have Started
That Exercise Program in the First Place

Suppose you read a book on healthy living and heart disease and decide to exercise, take aspirin, and eat foods that will lower your cholesterol level to 160. From the statistics in the book, you determine that this new lifestyle will decrease your chance of dying in the next five years from a 5% probability to a 2% probability (a 60 percent drop).

If you are like most people, your willpower and enthusiasm for healthy living will not last long and you will be back to your old bad habits. While sitting and eating a donut, you decide to crunch some numbers to see how much your probability of dying went up after abandoning your healthy ways.

Now you are really upset! Your healthy living dropped your probability of dying 60%, but now that you stopped, your probability of dying went up 150%! (Going from 2% to 5% is a 150% increase.)

Why don't you just say it went up three percentage points; that will make you feel a little better.

This example illustrates how expressing data with percents is ripe for distortion and manipulation. Obviously stopping your healthy living program will not raise your probability of dying more than starting the program lowered your risk.

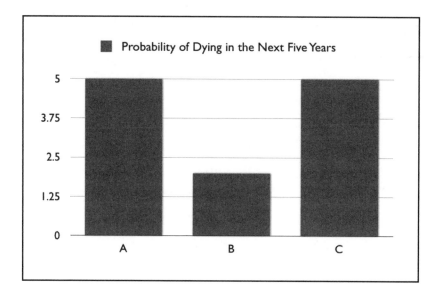

Before you started your healthy living program, you were at A. When you started your healthy living program, your probability of dying dropped to B, and when you quit, it went back up to C. As you can see from the bargraph, the amount of risk that went up and down is the same, but the percentage of change is very different.

A to B: Decrease of 60%   B to C: Increase of 150%

I'm just lucky I didn't stop smoking and hang gliding in addition to eating better, taking aspirin and exercising. Sure, I would have lowered my probability of dying by 90%, but if I resumed all my bad habits, I would increase my probability of dying by 900 percent!!

Look at the following examples of how selective use of the percentage of change can have a significant impact on how each situation is perceived.

The difference in fuel economy between a 20 miles per gallon pickup truck and a 50 miles per gallon Prius.

Choice 1:
If you go from a Prius to a pickup truck, your fuel economy will drop 60 percent.   (50 mpg to 20 mpg is a 60% drop.)

Choice 2:
If you go from a pickup truck to a Prius, your fuel economy will increase by 150 percent.   (20 mpg to 50 mpg is a 150% increase.)

A city added additional police officers and the crime rate dropped by 50% (say 1000 arrests declined to 500 arrests). Budget cuts may require that these officers be dropped from the force. Those who want to save the jobs of the police officers can present a warning about increasing crime to the public in one of two ways:

Choice 1:
The additional police officers on the force lowered crime by 50%.

Choice 2:
If these officers are let go, the statistics are very clear. Crime will probably increase by 100%.
(Going from 500 arrests to 1000 arrests is a 100% increase.)

Understanding Percents Could Save Your Life

Statins, which are used to lower cholesterol, are the most widely prescribed drugs in the United States and have been credited with saving tens of thousands of lives. The market for statins may expand even more after a recent clinical study showed that statins may be a life saver --- even for those without high cholesterol.

The study followed approximately 18,000 patients who did not have elevated cholesterol levels, but did have high CRP levels. (CRP refers to high-sensitivity C-reactive protein. High levels are indicative of inflammation in the body.) Half of the group was given a placebo, while the other half was given a powerful statin (Crestor).

The results of the planned five-year study were so dramatic that it was stopped after two years. The data monitoring board thought the benefits of the statin were so great that it would be unethical to continue the study because to delay would not allow patients in the placebo group to receive Crestor.

The group that took the statin had 31 heart attacks (.17%), while the control group had 68 patients who suffered heart attacks (.37%). The FDA's response to the clinical study was to approve Crestor for those without high cholesterol but with elevated inflammation and one additional risk factor such as high blood pressure or smoking. This new FDA action swelled the ranks of those approved for statins by 6 or 7 million.

Many in the medical field were excited about the potential of Crestor to save lives. They pointed to the 55% drop in heart attacks for the group that took Crestor and thought it would be foolish to deny a drug to a patient who met the criteria.

But not everyone was supportive of expanding the use of statins to millions of new people. Some of the critics thought that many doctors and consumers were misled by the widely reported 55% drop in heart attacks. They point to the very small .2 percentage points separating the statin group and the control group. One critic said: "That's statistically significant but not clinically significant".

Critics also were concerned about the rare, but real dangers of statins --- liver damage and muscle damage. They also pointed to new evidence that statins may increase the risk of type 2 diabetes by almost 10%.

If you are in the new group of 6 to 7 million who are now approved to take statins, which side do you listen to? Is the 55% drop in heart attacks the important number or is the miniscule .2 percentage points between .17% and .37% the better way to look at the data?

The data from the clinical trial shows that Crestor could have been responsible for preventing 37 heart attacks in the group of 9000 people who took the statin. This is approximately 4 prevented heart attacks per 1000 patients. If 6 million people who meet the criteria for taking Crestor actually do take the drug, then the data implies that 24,000 heart attacks could be prevented. The drop of 55% is a better interpretation of the data than the critics' "miniscule .2%".

Additional information about the Crestor study:

In August of 2010, the reliability of the Crestor study was starting to be questioned. Michel de Lorgeril, MD, of Joseph Fourier University and the National Center of Scientific Research in Grenoble, France and several co-authors did a reanalysis of the Crestor study. They questioned the conclusions of the study because:

• ".... close examination of the all-cause mortality curves shows that the curves were actually converging when the trial ended."

• The authors of the reanalysis concluded that total cardiovascular mortality was the same in the Crestor and placebo group.

• They also pointed out that nine of the fourteen authors of the Crestor study had financial relationships with the company that manufactures Crestor.

It looks like the "funding effect" may be at play here. It is probably wise to postpone judgment on the lifesaving power of Crestor until more information becomes available.

Discussion Questions

1) A state raised its sales tax from 4% to 5%. If you were an editor of a newspaper, would you report that the sales tax went up 1% or would you report that it went up 25%? (Moving from 4% to 5% is a 25% increase.)

2) Look at the results of the following (fictional) medical experiment:

  • There were 50,000 people in each group.

  • The group that took the heart medicine had .1% of its participants die of fatal heart attacks.

  • The control group (placebo) had .9% of its participants die of fatal heart attacks.

Which is the better way to report the results of the experiment?

(A) There was only a .8% difference in deaths between the two groups.

(B) There were 9 times the number of deaths in the group that did not receive the medicine.

Answers

1) Calling the sales tax increase 1% is somewhat deceptive. It would be much more informative to say that the sales tax went from 4% to 5% and that this will increase the state's sales tax revenue 25 percent.

You could also say that while the increase is only one penny per dollar, the taxpayer's burden would increase 25 percent.

2) These are dramatic results. There were only 50 deaths in the group that took the medication and 450 deaths in the control group. Choice B is a better choice.

400 lives saved for each 50,000 people taking a medication has the potential to save tens of thousands of lives. Aspirin, while not nearly as powerful as this fictitious medication, has been shown to dramatically cut the risk of heart attacks (20 to 30 percent drop).

CHAPTER SIXTEEN
Is Your Sample Fair?

I'm conducting a survey.
Should cats be allowed to live
indoors, or should they be
forced to live outside?

The results of my survey are very clear. 99.8% of the animals
that I interviewed felt very strongly that cats should not be
allowed in houses. I also found that 99.2% of the animals that I
surveyed felt that it was very important to keep cats
on a leash when they are outside.

The Importance of a Representative Sample

At the conclusion of a college reunion for the class of 1988, the president of the college, after surveying the 50 reunion attendees, announced with great pride that the average salary for the class of 1988 was a stunning $85,000!

Would it be fair for the college to mention the $85,000 average salary in its recruitment material?

> **A recent survey showed that one of our graduating classes has an average salary of $85,000! If you want a high salary, our college is the choice for you.**

When you try to determine whether the survey was fair, the most important question is whether the survey participants were a representative sample of the class of 1988. Was there as much selection bias as possible removed from the selection process?

An $85,000 annual salary is probably not an accurate representation of the typical salary of a graduate of the class of 1988 for the following reasons:

• Who is most likely to attend a class reunion? Is the group a fair sample of the class or is it more likely that those who have attained some career success will attend the reunion?

• Do you think Stan, who is homeless, will show up for the reunion to add his $2300 yearly income to the statistics?

• Were the participants honest?

• Was Bill Gates (or other income outliers) at the reunion to distort the mean?

• Did the president calculate the mean and median and use the higher number? (It is possible to have a median salary of $25,000 and at the same time have a mean salary of $85,000.)

The Roosevelt / Landon Election

Polling done before the presidential election of 1936 illustrates one of the best examples of why using a representative sample is so important.

The outcome of the presidential election of 1936 seemed to be easy to predict. Franklin Roosevelt was a charismatic leader who implemented programs through the New Deal that provided jobs for many voters during the Great Depression. In addition, he spoke to the American people on a regular basis during his popular radio fireside chats.

Roosevelt's Republican opponent, Alfred Landon, argued that Roosevelt's New Deal was not working. He also stated that Roosevelt was acting like a dictator and had become too powerful. Although Roosevelt was popular, Landon had many supporters throughout the country.

As election day approached, a magazine called the Literary Digest sent out 10 million postcard ballots in an effort to predict the winner. After two and a half million ballots were returned, the Literary Digest announced with great pride that their prediction was an overwhelming victory for Landon. Their polling showed that Landon would win with 57% of the vote to Roosevelt's 43%.

When the Literary Digest made its prediction, another pollster named George Gallup was amused because he had predicted almost the opposite outcome --- that Roosevelt would win with 56% of the vote. Gallup was very confident that he would be proven correct when the results of the election came in because he knew that there was a serious flaw in the way the Literary Digest conducted its survey.

The Literary Digest sent "ballots" to 10 million people whose names they found on lists of telephone and automobile owners. Gallup knew that a list of telephone and automobile owners was not an accurate cross-section of the American population. In 1936, only the wealthy and upper middle class owned cars and phones. The Literary Digest ignored much of the lower and middle classes.

Gallup only polled 50,000 people, but he used a representative sample. Because of this, Gallup's polling of a relatively small group of people was much more accurate than the Literary Digest's poll of 2.5 million potential voters.

The results of the 1936 election revealed that Roosevelt won with 62% of the vote. Gallup became a popular pollster after this election, while the Literary Digest lost its credibility as a polling organization.

When a Study's Participants Self-Select or the Sample Group is Biased, the Results are Almost Always Invalid

An article in Newsweek (November 13, 2000) showed how unreliable self-selected polling can be. Newsweek conducted an online vote during the run-up to the 2000 presidential election. People were asked to pick one of four responses to the following question:

Should Ralph Nader leave the presidential race?

• Yes, his liberal friends will regret it if Bush wins.

• Yes, we have a two-party system by design.

• No, a viable third party keeps Democrats honest.

• No, he is the only candidate worth voting for.

There were 51,717 responses:

• Yes, his liberal friends will regret it if Bush wins. **34%**

• Yes, we have a two-party system by design. **3%**

• No, a viable third party keeps Democrats honest. **54%**

• No, he is the only candidate worth voting for. **9%**

Because 9% said that Nader was the only candidate worth voting for, theoretically he should have received at least 9% of the vote in the 2000 election. He only received 3%. What happened?

This was a biased statistic because the sample wasn't randomly drawn from the population. The most likely explanation for the disparity between the polling results and the actual election was that a disproportionate number of Nader supporters participated in the poll, making him appear more viable as a candidate.

Another interesting study with flawed sampling (that Fox news falsely claimed was conducted by the New England Journal of Medicine) found that 46% of physicians would consider leaving their profession if the healthcare reform bill (2010) proposed by Democrats was passed. If this poll was representative of physicians throughout the nation, as the credible NEJM source suggests, this poll result would be staggering. The problem is that the poll was actually conducted by a medical recruitment firm (Medicus Firm). In fact, the only doctors sampled were those in the firm's database. This study was attributed to the New England Journal of Medicine numerous times, misrepresenting the methodology and therefore the credibility of this study.[31]

One of the best ways to understand why bias and/or self-selection invalidates most studies is to take a controversial issue, such as the wisdom of letting President Bush's tax cuts expire, and imagine the polling results from three distinct groups of participants.

- People who watch FOX news (This group would probably oppose letting the tax cuts expire.)

- People who watch news programs on MSNBC (This group would probably favor letting the Bush tax cuts expire.)

- A representative sample of all voters (This group's views about the wisdom of letting the tax cuts expire would probably fall somewhere between those of people who watch FOX news and those who watch news on MSNBC.)

The Importance of Double Blind, Randomized Clinical Trials

Before 1948, researchers would often get misleading results from their medical experiments because of bias. Sometimes the research consisted of doctors giving their patients a new medicine they thought would help them, followed by the same doctors making judgments about how effective the medicine was. Even when researchers split a group of patients into a control group and an experimental group, they may have consciously or unconsciously placed the least healthy patients into the group that did not receive the new medicine. This would make it much more likely that the medicine they were testing would appear to be effective.

The Control Group

The Experimental Group

This bias during the selection process rendered these experiments fairly useless. Any difference between the groups at the end of the experiment were most likely attributable to the health of the patients at the beginning of the experiment.

Another problem researchers had before 1948 was that they usually knew which patients received the experimental treatment, and which ones were in the control group. When the researchers tried to judge how much improvement occurred, they were consciously or unconsciously influenced by their knowledge of who was given the treatment.

If a doctor claimed that she had a treatment that could make all paralyzed patients walk normally again, she would not have to do a complicated research project to prove that her treatment worked. If paralyzed patients walked out of her office, that would be a strong indication of the effectiveness of her treatment. Unfortunately, the effectiveness of most medical treatments is not so obvious. For example, a heart medication that could prevent 10% of all heart attacks would save thousands of lives, but if a researcher gave it to only a few patients, the lifesaving benefits would not be apparent.

In 1948, British researchers devised a way to study medicines without letting bias ruin their results. In other words, they discovered a new way to find the truth. Their new method of experimenting was called randomized controlled trials.

What randomized controlled trials did was take judgment, hopes, dreams, and expectations out of medical research. It made the testing of medicine a true experimental science.

Randomized controlled trials required that patients be randomly assigned to either the test group or the control group. Additionally, researchers did not know which patients were getting the medicine and which ones were in the control group. At the end of the experiment, the two groups would be examined to see if the new medicine was effective or not (and also whether it was dangerous).

One of the first uses of randomized controlled trials was in 1948 when researchers decided to see if streptomycin was effective in the treatment of tuberculosis. One hundred seven patients were randomly assigned to two different groups. In this study, 52 people were assigned to a control group and 55 people were assigned to a group that received streptomycin.

Not only were the patients randomly assigned to the two groups, but the researchers did not know which patients were getting the medicine and which ones were in the control group. At the end of the experiment, the two groups were examined. The group that took streptomycin had a 7% mortality rate, while the control group had 27% of its patients die. The world now had a new and scientifically proven treatment for tuberculosis.

The Importance of the Size of the Test Group

The importance of the size of a test group cannot be stressed enough. An important study of statin's ability to prevent heart attacks (mentioned in chapter 15) would not have yielded the life-saving information it did if the sample size was too small.

The group that took a statin during the clinical trial had 31 heart attacks compared to 68 heart attacks in the group that was given a placebo. Each group consisted of approximately 9000 patients. The 55% reduction in heart attacks in the statin group was so dramatic that the study was stopped to give the control group the opportunity to benefit from the life-saving power of statins.

But what would have happened if the sample size was 100 per group instead of 9000? The statin group had one heart attack for every 290 participants, while the placebo group had one heart attack for every 132 participants. In a sample size of 100 for each group, it is very likely that both groups would have had no heart attacks. Another scenario is that each group had one heart attack or one group had no heart attacks while the other had one. The result would have been that little useful information would have come from such a small sample size.

Research does not always demand sample sizes in the thousands. If an experiment was done to determine the importance of using a parachute when jumping out of a plane, a very small test group, say 5 in the parachute group and 5 in the group that jumped without a parachute would provide fairly definitive proof of the importance of parachutes.

I wish I was in the group with the parachutes!!!

Our society has had the power of statistics available to us for hundreds of years; a power that when used properly, has the potential to save millions of lives. Unfortunately, there have been countless times when "statistical warnings" have been distorted, manipulated, and minimized. The result of this intellectual and moral failure has been the needless loss of millions of lives. Each member of our society must have the knowledge to understand the techniques that are used to distort and misuse numbers, know how to properly interpret statistical data, and have a skeptical mindset to assist in the process of ferreting out deception --- whether it is deliberate or unintentional.

Footnotes

1) *Doubt is Their Product*, David Michaels, page 148

2) *Physics for Future Presidents,* Richard A. Muller pages 282-286

3) *Doubt is Their Product*, David Michaels, page 150

4) *Doubt is Their Product*, David Michaels, page 148

5) *Supercrunchers,* Ian Ayres, page 2

6) *Supercrunchers,* Ian Ayres, page 105

7) *Supercrunchers,* Ian Ayers, page 7

8) *Doubt is Their Product*, David Michaels, page 9

9) *Doubt is Their Product,* David Michaels page 6

10) *Doubt is Their Product,* David Michaels page 8

11) www.tobacco.org/Documents/dd/ddfrankstatement.html

12) *The Funding Effect in Science and its Implications For the Judiciary,* Sheldon Krimsky, PH.D

13) *Side Effects*, Alison Bass, page 167

14) *Side Effects*, Alison Bass, pages 167-168

15) *Side Effects*, Alison Bass, page 172

16) *Side Effects*, Alison Bass, page 150

17) *Side Effects*, Alison Bass, page 183

18) *Side Effects*, Alison Bass, page 174

19) *The New York Review of Books*, Drug Companies & Doctors: A Story of Corruption, Marcia Angell

20) *Newsweek* 1/29/2010 The Depressing News About Antidepressants

21) *New York Times* Gardiner Harris and Eric Koli 06/10/05

22) *New York Times* Gardiner Harris and Eric Koli 06/10/05

23) *New York Times* Gardiner Harris and Eric Koli 06/10/05

24) Siegel: NPR 9/29/2009

25) Associated Press 4/7/2010

26) http://www.cbpp.org/cms/index.cfm?fa=view&id=1556

27) http://www.magnetictherapysales.com/testimonials.htm

28) http://www.harmony000.org/index.php?option=com_content&task=view&id=97&Itemid=271

29) http://www.ufomind.com/area51/testimonials/thieme.txt

30) *Damned Lies and Statistics*, Joel Best page 64

31) http://mediamatters.org/research/201003170046